International Climate Change Law and State Compliance

A solution to the problem of climate change requires close international cooperation and difficult reforms involving all states. Law has a clear role to play in that solution. What is not so clear is the role that law has played to date as a constraining factor on state conduct.

International Climate Change Law and State Compliance is an unprecedented treatment of the nature of climate change law and the compliance of states with that law. The book argues that the international climate change regime, in the twenty or so years it has been in existence, has developed certain normative rules of law, binding on states. State conduct under these rules is characterized by generally high compliance in areas where equity is not a major concern. There is, by contrast, low compliance in matters requiring a burden-sharing agreement among states to reduce global greenhouse gas emissions to a 'safe' level. The book argues that the substantive climate law presently in place must be further developed, through normative rules that bind states individually to top-down mitigation commitments. While a solution to the problem of climate change must take this form, the law's development in this direction is likely to be hesitant and slow.

The book is aimed at scholars and graduate students in environmental law, international law, and international relations.

Alexander Zahar is Senior Lecturer at Macquarie Law School, Macquarie University, Australia, where he teaches climate change and environmental law. His current research interests include the impact of international climate finance in Southeast Asia.

Routledge Advances in Climate Change Research

Local Climate Change and Society
Edited by M. A. Mohamed Salih

Water and Climate Change in Africa
Challenges and community initiatives in Durban, Maputo and Nairobi
Edited by Patricia E. Perkins

Post-2020 Climate Change Regime Formation
Edited by Suh-Yong Chung

How the World's Religions are Responding to Climate Change
Social scientific investigations
Edited By Robin Globus Veldman, Andrew Szasz and Randolph Haluza-DeLay

Climate Action Upsurge
The ethnography of climate movement politics
Stuart Rosewarne, James Goodman and Rebecca Pearse

Toward a Binding Climate Change Adaptation Regime
A proposed framework
Mizan R. Khan

Transport, Climate Change and the City
Robin Hickman and David Banister

Toward a New Climate Agreement
Todd L. Cherry, Jon Hovi and David M. McEvoy

The Anthropology of Climate Change
An integrated critical perspective
Hans A. Baer and Merrill Singer

Planning Across Borders in a Climate of Change
Wendy Steele, Tooran Alizadeh, Leila Eslami-Andargoli and Silvia Serrao-Neumann

Climate Change Adaptation in Africa
An historical ecology
Gufu Oba

Carbon Governance, Climate Change and Business Transformation
Edited by Adam G. Bumpus, James Tansey, Blas L. Pérez Henríquez and Chukwumerije Okereke

Knowledge Systems and Change in Climate Governance
Babette Never

Action Research for Climate Change Adaptation
Developing and applying knowledge for governance
Edited by Arwin van Buuren, Jasper Eshuis and Mathijs van Vliet

International Climate Change Law and State Compliance
Alexander Zahar

International Climate Change Law and State Compliance

Alexander Zahar

First published 2015
by Routledge
2 Park Square, Milton Park, Abingdon, Oxon, OX14 4RN

and by Routledge
711 Third Avenue, New York, NY 10017

Routledge is an imprint of the Taylor & Francis Group, an informa business

© 2015 Alexander Zahar

The right of Alexander Zahar to be identified as author of this work has been asserted by him in accordance with sections 77 and 78 of the Copyright, Designs and Patents Act 1988.

All rights reserved. No part of this book may be reprinted or reproduced or utilized in any form or by any electronic, mechanical, or other means, now known or hereafter invented, including photocopying and recording, or in any information storage or retrieval system, without permission in writing from the publishers.

Trademark notice: Product or corporate names may be trademarks or registered trademarks, and are used only for identification and explanation without intent to infringe.

British Library Cataloguing in Publication Data
A catalogue record for this book is available from the British Library

Library of Congress Cataloging-in-Publication Data
Zahar, Alexander, author.
  State compliance in international climate change law / Alexander Zahar.
    pages cm – (Routledge advances in climate change research)
  Includes bibliographical references and index.
  1. Climatic changes – Law and legislation. 2. Compliance.
  3. Global warming – Law and legislation. I. Title.
  K3585.5.Z34 2015
  344.04´633–dc23   2014027677

ISBN: 978-0-415-70817-3 (hbk)
ISBN: 978-1-315-88602-2 (ebk)

Typeset in Times New Roman
by HWA Text and Data Management, London

'Among the burgeoning legal literature on climate change issues, this book stands out—not only for its sober analysis of state compliance with the proclaimed basic multilateral commitments (perceived as "mitigation" and "accountability" rules), but especially for its critical long-term vision of an emergent "individualized" international law of the atmosphere.'

Peter H. Sand, University of Munich, Germany

'Much has been written about existing and proposed international climate agreements, but much less about whether what we have is actually working. Zahar's book fills an important gap. It probes the inconsistent compliance with the obligations under the Framework Convention on Climate Change and related agreements, it explores the usually feeble consequences of noncompliance, and it provides the lessons we should learn from this experience. This work will be invaluable to those working with the existing laws and those trying to craft better ones.'

Michael B. Gerrard, Columbia University, USA

'Compliance is a critical and often neglected part of the global climate regime. This book's most unique contribution is that it looks beyond individual emission reduction commitments and reporting obligations to consider a more complete range of individual and collective commitments under the existing climate regime to better understand the compliance challenge. This alone makes it essential reading for anyone interested in compliance with international environmental law.'

Meinhard Doelle, Dalhousie University, Canada

'Compliance is the elephant in the room in international climate law. This book offers a careful and timely analysis of why and when states choose (or not) to comply with international climate rules.'

Christina Voigt, University of Oslo, Norway

'A must for understanding the normative content of climate law. While providing a thorough analysis of procedural and substantive obligations imposed on states under the climate change legal regime, it sheds a new light on the interplay between individual and collective state compliance obligations.'

Massimiliano Montini, University of Siena, Italy

Contents

List of figures and tables   viii
List of abbreviations   ix
Glossary   xi

Introduction: why this book   1

1 Law, compliance, and the climate change regime   9

2 State compliance with rules on national reporting and assessment   28

3 Facilitation and enforcement of rules through the Kyoto Protocol's Compliance Committee   63

4 State compliance with emission-limitation obligations   86

5 Financial support for mitigation actions in developing countries   118

6 Climate law and 'optional' mitigation mechanisms   135

7 Compliance lessons from the climate change regime   165

Bibliography   175
Index   200

Figures and tables

Figures

2.1	Varieties of emission projections required for an Annex I party's national communication	32
4.1	Annual mean atmospheric concentration of CO_2 in ppm since 1990, as measured at Mauna Loa	108

Tables

1.1	Summary of treaty obligations under the international climate change regime	12
4.1	Quantified economy-wide emission-reduction targets, or pledges, by Annex I parties to the FCCC for the period 2013–2020	94
4.2	Nationally appropriate mitigation actions by non-Annex I parties to the FCCC for the period 2013–2020	97
4.3	Greenhouse gas emissions (Mt CO_2 eq.) of Annex I parties to the FCCC, including LULUCF	104
4.4	Greenhouse gas emissions (Mt CO_2 eq.), including LULUCF, from a sample of 47 non-Annex I parties to the FCCC	106
4.5	Greenhouse gas emissions (Mt CO_2 eq.) of Annex I parties to the Protocol in the first commitment period	112
4.6	Population 2000–2030 (thousands) and per-capita emissions (t CO_2 eq.), including LULUCF, in 12 Annex I parties to the FCCC	116
4.7	Population 2000–2030 (thousands) and per-capita emissions (t CO_2 eq.), including LULUCF, in 12 non-Annex I parties to the FCCC	117
6.1	Greenhouse gas net emissions and removals (Mt CO_2 eq.) from the LULUCF sector of the main countries in each of the world's three largest tropical rainforests	153

Abbreviations

AAU	Assigned Amount Unit: a unit issued pursuant to the provisions of Kyoto Protocol Decision 13/CMP.1. It is equal to one tonne of CO_2 eq.
ADP	Ad Hoc Working Group on the Durban Platform for Enhanced Action (FCCC)
AWG-KP	Ad Hoc Working Group on Further Commitments for Annex I Parties under the Kyoto Protocol
AWG-LCA	Ad Hoc Working Group on Long-term Cooperative Action under the Convention (FCCC)
CDM	Clean Development Mechanism, one of the three market (flexibility) mechanisms of the Kyoto Protocol
CER	Certified Emission Reduction: a unit issued pursuant to Article 12 of the Kyoto Protocol (Clean Development Mechanism) and the provisions of Kyoto Protocol Decision 3/CMP.1. It is equal to one tonne of CO_2 eq.
CMP	Conference of the Kyoto Protocol state parties
COP	Conference of the Parties to the FCCC
DNA	Designated National Authority (CDM)
DOE	Designated Operational Entity (CDM)
EIT	economy in transition
ERU	emission reduction unit: a unit issued pursuant to Article 6 of the Kyoto Protocol (Joint Implementation) and the provisions of Kyoto Protocol Decision 13/CMP.1. It is equal to one tonne of CO_2 eq.
EU ETS	European Union Emission Trading System
FCCC	United Nations Framework Convention on Climate Change, opened for signature in 1992, entered into force on 21 March 1994
GCF	Green Climate Fund, an operating entity of the FCCC's financial mechanism, created in 2010 by Decision 1/CMP.16
GEF	Global Environment Facility, the first operating entity of the FCCC's financial mechanism
Gt	gigatonne
IAR	International Assessment and Review (FCCC)
ICA	International Consultation and Analysis (FCCC)

IPCC	Intergovernmental Panel on Climate Change
JI	Joint Implementation, one of the three market (flexibility) mechanisms of the Kyoto Protocol
LDC	least-developed country
LULUCF	land use, land-use change, and forestry
Mt	megatonne
NAMA	nationally appropriate mitigation action (FCCC)
NAP	national adaptation plan (FCCC)
NAPA	national adaptation programme of action (FCCC)
NDC	nationally determined contribution (FCCC, post-2020 regime), and, as of 2014, 'Intended Nationally Determined Contribution'
NGO	non-governmental organization
NIR	national inventory report, an explanatory report submitted by Annex I parties together with their annual greenhouse gas inventory
ppm	parts per million
REDD	Reducing Emissions from Deforestation and Forest Degradation in developing countries, a mechanism under the FCCC
REDD-plus	An extended version of REDD, incorporating forest conservation, sustainable management of forests, and enhancement of forest carbon stocks in developing countries
RMU	removal unit: a unit issued pursuant to the provisions of Kyoto Protocol Decision 13/CMP.1 (emission reductions attributed to the land sector in Annex I parties), it is equal to one tonne of CO_2 eq.
SBI	Subsidiary Body for Implementation (FCCC)
SBSTA	Subsidiary Body for Scientific and Technological Advice (FCCC)

Glossary

2°C warming limit The upper limit of politically acceptable global mean temperature rise according to the 2009 Copenhagen Accord.

1.5°C warming limit A safer limit than the 2°C limit. It is referred to in the Copenhagen Accord as a possible alternative to the 2°C limit, if science (i.e. the IPCC) were to find that the higher limit is more dangerous than previously thought.

Annex B (Kyoto Protocol) List of emission-reduction obligations by Annex I parties to the Kyoto Protocol (in the form of a percentage increase/decrease from 1990 emissions) for the first commitment period under Protocol (2008–2012). For the second commitment period, an amended Annex B applies (see Decision 1/CMP.8).

Annex I party A party listed in Annex I to the FCCC, consisting of industrialized countries and countries in transition to a market economy (EITs).

baseline, or business as usual That which would have been observed, e.g. in terms of greenhouse gas emissions or another variable, in the absence of some form of intervention, e.g. a government policy or a CDM project; projection into the future of a historical trend.

climate change 'A change in the state of the climate that can be identified (e.g., by using statistical tests) by changes in the mean and/or the variability of its properties, and that persists for an extended period, typically decades or longer.'[1]

CO_2 equivalent Or CO_2 eq. A unit indicating that the global warming potential of a non-CO_2 greenhouse gas, or a mix of them (which may include CO_2), is expressed in terms of the quantity of CO_2 that would have the same warming effect.

Compliance Committee Created by the Kyoto Protocol, it makes determinations about state-party compliance with mandatory obligations on accounting and reporting of emissions and emission targets. It is divided into a Facilitative Branch and an Enforcement Branch.

---

1 Intergovernmental Panel on Climate Change (2014), *Climate Change 2014: Impacts, Adaptation, and Vulnerability: Working Group II Contribution to the Fifth Assessment Report of the Intergovernmental Panel on Climate Change: Summary for Policymakers*, p. 5.

Copenhagen Accord At COP 15 in 2009, informal negotiations in a group consisting of major economies and representatives of regional and other negotiating groups resulted in a political agreement known as the Copenhagen Accord. Over objections by a minority of states, the COP took 'note' of the Accord without adopting it. At COP 16, many of its features were adopted in regular decisions of the parties.

greenhouse gas The Kyoto Protocol controls the greenhouse gases listed in its Annex A. The list was extended in 2011 with the addition of nitrogen trifluoride (NF_3). The set of known greenhouse gases is larger than that controlled by the Kyoto Protocol. Some greenhouse gases are controlled by the Montreal Protocol to the Vienna Convention for the Protection of the Ozone Layer.

leakage, from a CDM project Greenhouse gas emissions unintentionally caused through the implementation of a CDM project.

leakage, specifically in forestry projects Where protection of forest carbon in one location causes carbon-emitting activities to shift to another location outside the project boundary.

non-Annex I party FCCC party not listed in Annex I to the Convention. Often used as equivalent to developing country.

Secretariat Administrative body of international public servants common to the FCCC and the Kyoto Protocol.

Introduction
Why this book

The periodic assessment reports of the Intergovernmental Panel on Climate Change, the fifth of which was published just prior to the completion of this book, are a testament to the maturity of the science of climate change.

The same cannot be said about the *law* of climate change. The first problem is the identification of climate change law, which is not a simple task. Its existence is often assumed in academic writings, but its laws, and the type of laws they are, are nowhere systematically spelled out. The text of the UN Framework Convention on Climate Change contains several quasi-ethical and quasi-legal principles, arrayed alongside practical rules to facilitate state cooperation. There is legal imperative in this treaty but it takes some effort to locate it.

Apart from an attempt to regulate the greenhouse gas emissions of a specific group of states up to the year 2000, the FCCC does not contain anything approaching a precise, enforceable, rule on the limitation of emissions at the state level. It is a glaring weakness, considering the treaty's objective to stabilize greenhouse gas emissions in the atmosphere. Still, there is enough climate change law in the treaty to enable a discussion of state compliance with the law to get started. The question that arises next is: Have states complied with the climate laws they created?

Once the current laws have been identified and the extent of compliance with them has been determined, it becomes easier to see how climate change law (or 'climate law' for short) still needs to be developed if the objectives of the climate change regime are to be met.

In this book, I limit the meaning of climate change law to law that proceeds from, and seeks to implement, a specific legal imperative, to the effect that greenhouse gases from human sources must be reduced, and eventually eliminated, with the aim of controlling climate change and the consequences that flow from the increase in average global surface temperature, which is the main impact of climate change. This might seem like a narrow definition. I return to its justification below.

In the chapters that follow, I argue that the FCCC has created new international law that imposes two main requirements on states. First, they must report their greenhouse gas emissions in a detailed, transparent, and reviewable manner in regular inventories (the 'accountable reporting' rule). Second, it obliges states collectively to prevent 'dangerous' climate change (the 'general mitigation' or 'prevention' rule). So well-grounded are these obligations in the community of states that a plausible argument could be made that both of them now have achieved the status of customary international law, directing state conduct and restricting what states are free to do as a matter of universal consensus and essentially independently of the FCCC itself. However, for reasons that I will set out in due course, state conduct does not fully support the CIL argument. Still, the two laws I have just alluded to are by now very well grounded in universally subscribed-to treaty law and normative justifications.

The FCCC's offshoot treaty, the Kyoto Protocol, is more concrete and prescriptive than the FCCC about mitigation targets for individual states. It is also narrowly targeted at developed states. These so-called Annex I parties to the Kyoto Protocol[1] have complied with its rules—more or less. However, the Protocol has not developed any new international climate law, nor is it, conceptually, an advance on the FCCC. It perpetuates bottom-up (domestically determined) instead of top-down (scientifically determined) emission targets. While it still has some life in it, it is already considered doomed. It will limp on until 2020, after which it will be retired. The Protocol's contribution to the development of climate change law does not go beyond its affirmation of the rule well established under the FCCC that states must report their greenhouse gas emissions fully and accountably.

A new global deal, whose form and content are still indistinct, is expected to be implemented as of 2020/2021, after being agreed to by the end of 2015. There is no certainty, of course, that a post-2020 agreement will be agreed to in 2015, or, if agreed to, that it will be ratified by all states with high emissions. Such 'expectations' have come to naught before. The legal point I am making is that presently there is no clarity about whether the legal force of the next international climate change regime will venture beyond the FCCC's two well-established laws, namely accountable reporting (expounded in Chapter 2) and general mitigation or prevention (Chapter 4).[2] It is remarkable that the category of international climate law has been so neglected in the literature that these two laws I have identified, which certainly make up most of the category's substance, have not so far been articulated. When the International Law Association, an NGO, undertook recently to codify the 'legal principles relating to climate change', it managed only to apply pre-existing legal principles to the new facts of climate change, thus perpetuating a misapprehension that this is already a rich field of international law, while at the same time neglecting to mention the two rules of law that the international climate change regime has in fact managed to establish.[3]

State-level action on the mitigation of greenhouse gas emissions is increasingly a common practice. Many countries, both developed and developing, have pledged emission reductions, initially under the Copenhagen Accord in 2009 and

---

1 Annex I parties to the Kyoto Protocol are a subset of Annex I parties to the FCCC.
2 At other times I refer to the former as the 'transparency' rule.
3 Shinya Murase and Lavanya Rajamani (et al.) (2014), *Legal Principles Relating to Climate Change: Report and Draft Declaration for Consideration at the 2014 Washington Conference.*

subsequently under an FCCC plan covering the 2013–2020 period. Such action at the state level might be taken as evidence of an unstated general rule of law at work, namely that any country with relatively high or steadily rising emissions is obliged as a matter of international law to commit itself to emission reductions, or at least emission limits, now and into the future. Unfortunately, there is little evidence to support the inference. Developing countries (the vast majority of non-Annex I parties to the FCCC)[4] have emphasized that their mitigation pledges (or, to use the official term, 'nationally appropriate mitigation actions') are not only self-determined and non-binding (for this much is also true of Annex I party pledges), but that they are proffered pursuant to no legal obligation. There being no such obligation in their view, many developing countries have not pledged any mitigation actions at all, or have only pledged weak or symbolic ones. As for the Annex I parties to the FCCC, essentially all of them now subscribe to the United States-led position that no obligation to limit emissions applies to them alone as law, because the FCCC's prevention rule necessarily encompasses all 'major emitters', including several fast-developing economies in the non-Annex I group, without whose cooperation dangerous climate change cannot be avoided. In this fashion, the legal upshot of state-level action is played down by the two sides. Thus, an inference to a new international norm on state mitigation representing an advance on the already established 'prevention' law becomes very difficult to draw.

State-level mitigation pledges for the 2013–2020 period have not, moreover, all been of the same kind. The plurality in their form is a further obstacle to any argument that a general rule of law underlies state mitigation action during this period. A country that promises to reduce its emissions from a 'business as usual' level essentially has three options. First, it may promise to reduce the 'emission intensity' of its economy (fewer emissions per dollar of GDP), leading to a reduction in emissions from a hypothetical future trajectory where emission intensity remains the same (or where it changes at a business-as-usual rate). For a strongly growing economy, like China, it is a scenario that allows for emissions to continue to rise in absolute terms. Its virtue, from a verificationist point of view, is that emission intensity is a quantity that can be measured, and thus the country can be held accountable to its target.

A second form of mitigation pledge is one that promises a reduction from business-as-usual emissions without specifying the indicator to whose adjustment the mitigation effort will be directed. For example, a country may promise to institute a policy to incentivize energy efficiency or to afforest land. The weakness in this form of pledge is that it does not commit a country to a particular quantifiable variable (or set of them) that is strongly correlated with, and thus a good indicator of, country-wide emissions. If the country's business-as-usual trajectory is accepted by the international community as validly determined, and the promised policies are implemented by the government, a continued increase in

---

4 The terms 'developing country' and 'non-Annex I party' are often used interchangeably. However, not all non-Annex I parties consider themselves developing countries.

the country's emissions could be presented as a reduction from business as usual even if no proof is offered apart from the implementation of the policies. Some countries will be able to build a case for mitigation more persuasively than others using this model. Most will be unable to quantify the supposed 'reduction' with any degree of accuracy, or at all.

The third kind of pledge is the most precise and reliable. It consists of an absolute reduction in emissions compared with a past, measurable emission baseline. It does not allow for territorial emissions to rise (unless trading in emission allowances is permitted, in which case a state could pay to increase its limits), and provides an objective, straightforward, indicator of success. A variation to this kind of pledge might be arranged by a country whose population grows by a significant amount each year due, say, to immigration.[5] Such a country might prefer to express its pledge as a reduction in per-capita emissions compared with a past date. If the promise is kept, the country will have reduced its greenhouse gas emissions with respect to the baseline population despite an increase in absolute emissions in its territory.

In short, mitigation commitments can range from the merely arguable to the provably real.

While a small number of developing countries with pledges for the 2013–2020 period have chosen emission-intensity reductions, most have pledged a reduction from business-as-usual emissions without committing to a key indicator that would serve as proof of their mitigation actions.[6] Either way, the emissions of developing countries are expected to continue to grow in absolute terms through to 2020 (and probably beyond). They might grow more slowly because of the pledges, or their growth might be unaffected by the pledges—it will be difficult to know.

Annex I party pledges, by contrast, are all of the third kind: they are absolute reductions rather than, at best, managed growth. Several Annex I party pledges are conditional, leaving open a course of more ambitious reductions to follow but only if other Annex I parties do the same. Even within this form, then, there is variation in state action, from the real to the potential. In emphasizing that all Annex I parties have pledged absolute reductions by 2020, I am not meaning to single out this group of countries for any special praise. In fact, there is much to be said about their lack of ambition and leadership in climate change matters. My only point is that countries go about mitigation in substantially different ways, and about half are doing nothing at all about it, so we would be searching in vain for a norm underlying state conduct that we do not already know about from the text of the FCCC, notwithstanding that it is more than 20 years old.

---

5 Several examples could be given. The major net receivers of international migrants during 2010–2050 are projected to be the United States (1,000,000 annually), Canada (205,000), the United Kingdom (172,500), Australia (150,000), Italy (131,250), Russia (127,500), France (106,250), and Spain (102,500): United Nations Department of Economic and Social Affairs (2013), *World Population Prospects: The 2012 Revision. Volume I: Comprehensive Tables*, ST/ESA/SER.A/336, p. xxi.

6 See Table 4.2 in Chapter 4.

The differentiations I have outlined leave the Copenhagen track of the FCCC no more advanced, legally, than the Kyoto Protocol track, which binds only Annex I parties to absolute emission reductions (albeit allowing a few of them to make slight emission increases). Under the Copenhagen track, it is still only Annex I parties that have agreed to commit to actions to reduce, and not simply to moderate, the growth of their greenhouse gas emissions.[7]

Even though this book does not present a survey of domestic climate change law (others have done so), it may be inferred from what I have said so far that there is no rich body of climate law subsisting in domestic jurisdictions, waiting, as it were, to be recognized as universal legal practice and as a source for the further development of international law. The extent of climate law at the domestic level has indeed been exaggerated in certain accounts. There has been a well-intentioned tendency to count, not just every legislative provision but every policy initiative or strategy with a bearing on climate change as an instance of climate law.[8] Some of the domestic action that gets lumped into this mix is about adaptation (which I do not count as climate law). Other elements of it that are purportedly about mitigation do not necessarily reduce emissions, or they are mere window-dressing, or have seen their political support collapse from one year to the next (most recently in Australia) causing a return to the old status quo of no climate law, or they have an insecure political or legislative basis because there is no normative basis beneath that. The law of the European Union has undoubtedly had a moderating influence on the emissions of its member states, and important jurisdictions such as the United Kingdom have enacted control levers that could be put to good use when (or if) the time comes for decisive action. In general, however, and notwithstanding optimistic announcements about a proliferation of domestic climate law, municipal legal norms for the attenuation of climate change through compulsory emission limits remain tenuous.[9]

My definition of climate law narrows the concept to a kind of mitigation law. This I reluctantly accept, because there is no plausible alternative. The prevention rule is at the heart of climate law as it has evolved to date. The rule on accountable reporting exists essentially in order to facilitate the FCCC's 'ultimate objective' as set out in article 2 of the treaty—that is, mitigation. There are risks to tacking other rules onto this modest body of law in order to say that it is also about

---

7   There is an inconsequential exception to this general statement. It involves four very small developing countries: see Table 4.2 in Chapter 4.
8   GLOBE is the cheerleader on state lawmaking on climate change. Yet this NGO's compilations—see, most recently, Michal Nachmany et al. (2014), *The GLOBE Climate Legislation Study: A Review of Climate Change Legislation in 66 Countries (4th edn)*—conflate law, on the one hand, with policy, planning, and good intentions, on the other. The same is true of Navroz K. Dubash et al. (2013), 'Developments in National Climate Change Mitigation Legislation and Strategy', 13 *Climate Policy* 649.
9   For the sceptical view along these lines, arising from, but not limited to, the legal rhetoric of a developing country, see Michael Kidd and Ed Couzens (2013), 'Climate Change Responses in South Africa', in *Climate Change and the Law*, ed. Erkki J. Hollo, Kati Kulovesi, and Michael Mehling.

adaptation. Just as the sum total of equine regulation does not bring into being a 'horse law', so laws and litigation and other forms of legal intervention that relate to climate change do not necessarily amount to climate law.[10] Thus, a coastal-protection law need not be counted as part of climate law, even if it was enacted specifically to defend against the impacts of climate change. Such a law does not necessarily contribute to the prevention of climate change and it might even exacerbate the problem (if, for example, it calls for works that lead to the release of even more greenhouse gases). Where a law is limited to the management of the impacts of climate change, an accurate term for it would be 'law for adaptation', distinguishing it from law that relates to the management of climate change itself and to the core purpose of the FCCC. Even a coastal-protection law that contains an element of mitigation, such as a plan to reforest coastal areas to reduce erosion, is not necessarily, in this aspect, a kind of climate law, for the state's larger plan might be to surrender inland areas to deforestation in exchange, leaving no net gain in carbon sinks. In the search for a specialized climate law one must proceed cautiously, lest rules unrelated to the control of climate change or lacking a secure normative basis, are included.

In sum, the most that may be inferred from domestic mitigation practice to date is that some states seem bound to take *some* action to limit their emissions. It is not an enlightening inference, for it does not take us beyond the original FCCC commitment (the prevention or general mitigation rule). No matter which path we try, I intend to show, the conclusion is that substantive international climate law has not advanced beyond the 1992 treaty.

The post-2020 agreement is likely to commit more countries to emission reductions, but until the language of that agreement is finalized, we will not know whether it represents a progressive development of international law toward a universal rule of compulsory and equitable action on mitigation tied to a global 'emission budget' calculated to avoid dangerous warming, or whether it is just a reiteration of the FCCC/Kyoto Protocol idea that it is for each state to decide, on social and economic grounds of its own choosing, what level of mitigation effort, if any, to commit to.

With state responsibility for mitigation still a fraught topic, the international regime has advanced on an easier front, creating a complex of procedural rules on the objective reporting of emissions and government policies to control them. 'Accountable reporting' is a rule of climate law that enjoys normative force because it is responsible for some of the preconditions needed for ambitious mitigation action by states. This book devotes considerable space to the examination of accountable reporting.

There is yet another rule of international climate law that might at first sight appear distinct from the other two and of such a nature as to be categorized as procedural. Annex I parties are obliged by the FCCC to provide financial support to developing countries, to the extent it is needed, to enable them to meet the

---

10 See J. B. Ruhl and James Salzman (2013), 'Climate Change Meets the Law of the Horse', 62 (4) *Duke Law Journal* 975.

treaty's objectives on reporting, mitigation, and adaptation. The main difficulty with viewing this rule as procedural is that, given that the FCCC's objective (article 2 of the treaty) is to reduce global emissions, the law on financial support, to the extent that it serves the FCCC's objective by supporting mitigation in developing countries, is coterminous with the prevention or general mitigation rule. Financial support is thus more accurately characterized as a substantive rather than procedural requirement, and it goes hand-in-hand with the prevention rule. The rule on climate finance is in the same situation as the prevention rule: while Annex I parties have been providing *some* financial support for mitigation in developing countries, in the absence of a global emission budget informed by the need to avoid 'dangerous' change and an agreed division of responsibility for financial contributions to assist countries collectively not to exceed the budget, international financial support for mitigation has been ad hoc.

The reporting obligation is, as I have said, a well-developed area of procedural climate law that enjoys a high level of compliance and a steadily expanding practice. However, where the rule applies to the reporting of climate finance, state compliance is less impressive. Not only do Annex I parties provide ad hoc amounts of funding for mitigation and are thus non-compliant on the substantive side, their reporting on financial support suffers from a lack of transparency. In this particular area of practice, states are non-compliant with the rule on accountable reporting (as explained in Chapter 5).

Some rules of the climate change regime appear to be merely 'optional', in the sense that they bind states only when they have elected to opt in to certain mechanisms. The book discusses the CDM and REDD under the optional category (Chapter 7). The point of this analysis is to show that state non-compliance with the rules of optional mechanisms could undermine the implementation of the climate change regime's basic laws.

A succinct response to the title of this Introduction, 'Why this book', is that the content of international climate law currently lacks definition. Once better defined, the discussion can be turned to other interesting questions, such as whether states comply with the law that does exist, what compliance mechanisms, if any, oversee the implementation of their obligations, whether states need to develop more climate law, and, if so, in which specific areas must they do so, using which, if any, compliance mechanisms. The provisions of the FCCC and the Kyoto Protocol are an obvious place to start the inquiry in this book. The two treaties appear to impose a range of obligations on states. The extent to which these obligations have been discharged, or not, is far from being a simple matter of fact. As I have indicated, it is not always possible to discern what the legal obligation is or what the state practice is. The book focuses on the 1997–2014 period: from the time when the FCCC came into force until the time of writing.

Should we be concerned about climate law at all if it is still in a poor state after more than 20 years? An optimist who has no time for law might consider that the science of climate change is well advanced and scientific advice will eventually gain the upper hand. On this view, government decision-makers will eventually take strong mitigation action without having to pass through the stage of recognizing that there exists a legal imperative to do so. Progress in getting climate change under control, according to the optimist, will come from strong policies and political cooperation, informed directly by science, and the state of the law will not be critical to the endgame, in which humanity manages finally to gain control. The IPCC's reports might well impart this flavour of scientific positivism. The problem with the notion that it all boils down to science and policy, not law, is that without a shared sense of legal obligation to provide a system of values, coordinated action against climate change on a global scale is unlikely to evolve fast enough in the direction of strong mitigation action. As the IPCC's fifth assessment report makes clear, scientists agree that emissions must be reduced sooner rather than later if we are to avoid catastrophic environmental change.[11] Having in place laws that one complies with out of a sense of necessity and justice, while there is still time to avoid a climatic crisis, seems a necessary element of the solution.[12] Politicians are unlikely to listen to scientists unless the law says that they must.[13]

---

11 See Section 4.2.1 in Chapter 4.
12 Gavin A. Schmidt (2014), 'On Scientists and Advocacy', 344 *Science* 256.
13 On the mentality of politicians in this respect, see Michael Howlett (2014), 'Why Are Policy Innovations Rare and So Often Negative? Blame-Avoidance and Problem-Denial in Climate Change Policy-Making', *Global Environmental Change*, in press.

1 Law, compliance, and the climate change regime

1.1 A low profile for law

The IPCC's fifth assessment report, at close to six thousand pages, is an encyclopaedia on the Earth system—its future as much as its past. The work is the culmination of an impressively methodical and coherent response of the natural sciences, along with certain disciplines of the humanities, to the problem of climate change. The IPCC's report gives prominence to several areas of inquiry. Questions of law and the work of law academics are not among them. Economists and policy specialists are given substantial coverage, by contrast.

In a few places where the report mentions the law as a subject of academic inquiry rather than as a mere instrument of policy, the IPCC's implication is that legal research has not provided useful answers: 'Research has not resolved whether or under what circumstances a more binding agreement elicits more effective national policy.'[1] An exception to the general attitude might be this statement from the report: 'Because greater legal bindingness implies greater costs of violation, states may prefer more legally binding agreements to embody less ambitious commitments, and may be willing to accept more ambitious commitments when they are less legally binding.'[2] However, the statement seems a priori: a deduction from common sense rather than a true discovery. It too carries a negative connotation about the law in its suggestion that a commitment with legal force could retard ambitious action against climate change.

If the IPCC's reports were taken as the measure of relevant enquiry on climate change, we would have to concede that law academics have yet to secure a disciplinary space for themselves or have their insights recognized outside their own field. Alternatively, their contributions or failings are being attributed to international relations, economics, or policy analysis—a classification that in many cases is deserved because the legal content of articles on climate change issues published in law journals is often minimal. Law—that 'system of rules that a particular country or community recognizes as regulating the actions of its

---

1 Intergovernmental Panel on Climate Change (2014), *Climate Change 2014: Mitigation of Climate Change: Working Group III Contribution to the Fifth Assessment Report of the Intergovernmental Panel on Climate Change (Final Draft)*, ch. 13, p. 26.
2 Ibid., ch. 13, p. 27.

members and may enforce by the imposition of penalties'[3]—seems to have played a subservient role in the overall response to climate change, even as many legal scholars have engaged professionally and at length with the climate change field.

Niggling concerns about the role of academic law in the conversation on climate change invite us to consider the extent of the substantive climate change law presently in place and its contribution to the conduct of states in the international community. In Mayer's words, the search is for a 'climate law in a narrow sense—a set of norms forming a special legal regime, having its own object and purposes and its own doctrine'.[4] It is also an inquiry into the possibility that climate law is much less than what some legal academics have assumed,[5] both in substance and as a guide to states. It could explain the law's poor showing in the IPCC reports.

Questions about the existence of climate law and the compliance of states with the requirements of the climate change regime are linked, as suggested by this book's title. The IPCC's reports may be silent on the law, but they are vocal about the fact that state action against climate change is insufficient and even reckless. The two conditions could be related. An absence of climate law would be relevant to an explanation of state conduct found to be unambitious or incoherent, just as an acknowledged and binding law would be relevant to an explanation of state conduct found to comply with the regime's objectives. For explanatory reasons it is important to know whether the state conduct criticized by the IPCC occurs in a context of laws or in a context not defined by clear legal obligation.[6]

Law is often merely facilitative. But it can also be a force in its own right. Where it has such force, it is said to have a normative quality (normativity). Beyerlin calls it 'the capacity to directly or indirectly steer the behaviour of its addressees', and Porter calls it a law that has 'purchase on a community'.[7] Norms are 'prescriptions for action in situations of choice, carrying a sense of obligation,

---

3 *Oxford English Dictionary*.
4 Benoît Mayer (2013), 'Climate Change and International Law in the Grim Days', 24 (3) *European Journal of International Law* 947, p. 954.
5 See, e.g., Catherine Redgwell (2012), 'Climate Change and International Environmental Law', in *International Law in the Era of Climate Change*, ed. Rosemary Rayfuse and Shirley V. Scott, p. 118 (the FCCC and the Kyoto Protocol 'have significantly enriched the corpus of general international environmental law rules and principles').
6 The general point has been noted by Kal Raustiala and Anne-Marie Slaughter (2002), 'International Law, International Relations and Compliance', in *Handbook of International Relations*, ed. Walter Carlsnaes, Thomas Risse, and Beth A. Simmons, p. 538 ('Law and compliance are conceptually linked because law explicitly aims to produce compliance with its rules').
7 Ulrich Beyerlin (2007), 'Different Types of Norms in International Environmental Law: Policies, Principles, and Rules', in *The Oxford Handbook of International Environmental Law*, ed. Daniel Bodansky, Jutta Brunnée, and Ellen Hey, p. 428; and Jean Porter (2007), 'Custom, Ordinance and Natural Right in Gratian's *Decretum*', in *The Nature of Customary Law*, ed. Amanda Perreau-Saussine and James Bernard Murphy, p. 100.

a sense that they ought to be followed'.⁸ Law of this kind arises from principles of justice, morality, or other first principles.⁹ It has been called a 'legally binding norm' or a 'fixed norm'.¹⁰ Applied to the international sphere, normative law is capable of constraining and shaping the behaviour of states. It has 'causal significance' in relation to state behaviour.¹¹ Normative law is different from 'facilitative law' (the implementation arm of policy) which is instrumentalist and subordinate to politics, and different again from soft law, which guides but does not compel.

One could illustrate the special quality of normative law by drawing from the arena of human rights law. There, solidifying legal norms in the second half of the twentieth century compelled legislatures to pass laws to regulate both state and individual conduct and penalize the abuse of 'human rights', for reasons not only of government policy but of (newly recognized) legal imperative. Once enacted, the protection of human rights became an institution to be reckoned with, and because it was backed by independent norms it ran a low risk of repeal. Similarly, large tracts of pollution law could not, nowadays, conceivably be repealed. They attach to ingrained norms. At the other end are technical laws that merely implement policy and have no value outside of their instrumentalist, facilitative, function.¹²

Where a legal 'system of rules' develops in an area of conduct that is seen as important to national well-being, a legal specialization tends to form around it. Human rights lawyers, practising and academic, work with such a system of rules. Human rights law is underpinned by a system of justice, of right versus wrong. The comparison with human rights law makes climate change law look abject. While it has gained some ground since the 1990s, it has yet to give rise to a *system* of rules. The few rules that it can lay claim to are fragmented and weak. There is some potential for this to change, although the day when climate law becomes a regular legal specialization seems far off.

---

8   Abram Chayes, Antonia Handler Chayes, and Ronald B. Mitchell (1998), 'Managing Compliance: A Comparative Perspective', in *Engaging Countries: Strengthening Compliance with International Environmental Accords*, ed. Edith Brown Weiss and Harold K. Jacobson, p. 42.
9   Thomas M. Franck (1995), *Fairness in International Law and Institutions*; and Harold Hongju Koh (1997), 'Why Do Nations Obey International Law?', 106 *Yale Law Journal* 2599, p. 2628.
10 Ulrich Beyerlin (2007), 'Norms in International Environmental Law', p. 435.
11 William Bradford (2005), 'International Legal Compliance: Surveying the Field', 36 (2) *Georgetown Journal of International Law* 495, p. 495.
12 In saying that some laws are more tenacious than others because of their strong or common moral foundation, whereas other laws are merely technical and can be easily amended or repealed, I do not engage with, or take a position on, the classical debate between positivism and non-positivism. It could be that technical laws have a contested or indistinct moral foundation, or that they have no such foundation at all. My argument relies on a factual observation and is not a comment on the nature of law.

12 Law, compliance and the climate change regime

Table 1.1 Summary of treaty obligations under the international climate change regime. The letters in the table are explained immediately below it.

	Group	Pre-2008	2008–2012	2013–2020	Post-2020
FCCC	Annex I	A.		B.	C.
	Developing	D.		E.	F.
Kyoto	Annex I	G. C.C.→	H.	I.	K
	Developing	J.			

A. Annex I parties to the FCCC must implement and report on mitigation measures as well as calculate and report on their emissions. Annex II parties must fund certain costs of developing countries. No mitigation targets are specified in the treaty beyond the year 2000, by which date Annex I parties were to return their emissions to 1990 levels.
B. Quantified emission reductions are pledged by Annex I parties against historical baselines. The pledges are self-determined and not legally binding. There is biannual reporting on the pledges, to be reviewed under the IAR (International Assessment and Review) process.
C. Agreement on the post-2020 regime is scheduled for late 2015 (a highly optimistic deadline). The broken line between C and F indicates that the FCCC's original, hard distinction between Annex I and non-Annex I parties might not survive the transition to the new agreement.
D. Mitigation and adaptation measures are expected by developing countries, as is reporting on emissions and other matters, but no actions are compulsory and all actions are formally conditional on the availability of endogenous capacity and financial support from the Annex I group. Note that the general mitigation or prevention rule (to avoid dangerous climate change) applies to all FCCC parties collectively.
E. Mitigation pledges are mostly in the form of reductions against business-as-usual forecasts, are voluntary and non-binding, and are reviewed under the ICA (International Consultation and Analysis) process.
F. Same as C.
G. The main obligation prior to 2008 was for Annex I parties to set up 'national systems', calculate base-year emissions (using the same reference year as for the FCCC, i.e. 1990), and agree on their 'assigned amount' of emissions for the first commitment period. The Compliance Committee of the Kyoto Protocol (labelled C.C. in the table) became operational in 2006.
H. *First commitment period:* Quantified emission-reduction obligations were in force over this period for most Annex I parties (the United States was the main exception). Annex I parties had an option to engage in various forms of emission trading. When these were chosen, they imposed additional sets of obligations. Canada withdrew from the Protocol just prior to the close of this period.
I. *Second commitment period:* Same as H, except that the quantified amounts are different, and three more Annex I parties (Russia, Japan, and New Zealand) refused to participate.
J. No new obligations for developing parties were imposed by the Protocol. Technically, the Facilitative Branch of the Compliance Committee may also assist non-Annex I parties with their voluntary actions, but this has not eventuated in practice.
K. No one expects a third commitment period under the Kyoto Protocol. Without a commitment period following 31 December 2020, obligations and compliance under the Protocol will cease to be live issues and the treaty will fall into abeyance.

1.2 International climate law

The structure of obligations under the international climate change regime is shown in Table 1.1. Two treaties, the FCCC and the Kyoto Protocol, each with different obligations for developed and developing countries, have expanded their operations and enriched their terminology over time, through decisions of the state parties and the (much rarer) adoption of amendments. The most salient transition years were 2008 for the Kyoto Protocol and 2013 and 2020 for both treaties. Of the two treaty regimes, the obligations under the FCCC have changed the most, and are expected to continue to change. The FCCC is a much more general and accommodating regime than the Kyoto Protocol.

The normative law that has emerged from the two treaties is discussed in detail in Chapters 2 to 5.

From its very beginning, the international climate change regime did not seek to reverse global warming. It accepted that some warming was inevitable. According to the FCCC, states are to limit their emissions to avoid 'dangerous' warming.[13] With time, the IPCC offered a quantification of the upper limit: 'a 2°C increase [is] an upper limit beyond which the risks of grave damage to ecosystems, and of non-linear responses, are expected to increase rapidly'.[14] That limit was adopted by the FCCC: 'deep cuts in global greenhouse gas emissions are required according to science ... so as to hold the increase in global average temperature below 2°C above pre-industrial levels'. FCCC parties are required to 'take urgent action to meet this long-term goal ... on the basis of equity'.[15]

The imperative expressed in the last two sentences may be considered the foundation of international climate law. It specifies the action required of states (deep cuts in emissions), a method by which to calculate the necessary and sufficient action by states (avoidance of 2°C warming), and a response time line (urgent action, which suggests immediate action). It does not specify how the action is to be shared among states, beyond the requirement that it must be done equitably. The normative content of the law could hardly be more potent: its breach could lead to 'grave damage to ecosystems'. Ecosystems support human welfare, so acting to save them is a matter of necessity, perhaps moral necessity, not choice.

Climate law thus has a moral, existential foundation which rivals that of any other area of the law. What it lacks is elaboration in the direction of greater specification. Until states can agree on how to implement the general law, and through that agreement create new, concrete obligations, cuts in emissions will be gestures of good will and always fall below what the general law requires.

The FCCC's foundational rule helps us to distinguish between rules of law that are closely or remotely related to it. That is, it gives us a workable definition of climate law that is not so broad as to tempt us to assemble a kind of 'horse law'. An obligation upon a state to limit greenhouse gas emissions in its territory to a specified amount below (or, where justified by its circumstances, above) a historical reference year may be called a 'core-outcome obligation' because it directly addresses the objective of the FCCC to reduce anthropogenic emissions to safe levels. Even if the obligation is such that it applies only to a minority of

---

13 FCCC, art. 2.
14 Intergovernmental Panel on Climate Change (2007), *Climate Change 2007: Mitigation: Contribution of Working Group III to the Fourth Assessment Report of the Intergovernmental Panel on Climate Change*, p. 99.
15 FCCC (2011), *Decision 2/CP.17, Outcome of the Work of the Ad Hoc Working Group on Long-Term Cooperative Action under the Convention*, FCCC/CP/2011/9/Add.1, part II.A, preamble.

states that cannot by their own combined effort reduce global emissions to safe levels because of their low proportional contribution, it is still close to the core of the international climate change law because its effect is to limit emissions, even if not by a globally meaningful amount. This *kind* of rule has also been called a primary rule[16] or a core treaty goal.[17]

The FCCC itself contains one, and only one, core-outcome rule, now expired. I discuss it in Chapter 4, where I refer to it as the specific mitigation rule (to be contrasted with the general mitigation rule discussed above). The climate change regime's most famous core-outcome rule is the Kyoto Protocol's capping of the emissions of (most) Annex I parties over two commitment periods (2008–2020). The rule has this form: 'The country shall limit its emissions to a fixed amount above or below a designated reference year.' In the first commitment period, the rule obliged Annex I parties to keep their averaged annual emissions for 2008–2012 within a certain percentage of their 1990 emissions. The rule is directly connected to the FCCC's foundational law as it operates to mitigate the causes of climate change in several developed countries. According to the definition I have presented, it is climate law. However, the Protocol rule does not reduce global emissions to safe levels. It is a step removed from the innermost core of climate law. The difference is significant, because the normative hold of the FCCC's foundational law derives entirely from its objective to prevent a dangerous alteration of the Earth system.

The Protocol rule is a creature of treaty law, cut down by qualifications. By 2014 the rule had occupied the climate-law pedestal for about seven years but was not generally held in high regard. The number of Annex I parties rejecting the treaty had risen to five.[18] Only a handful of countries had formally accepted the Doha Amendment to the Kyoto Protocol which put in place the mechanisms for the second commitment period. It is just possible that the amendment will never formally enter into force.[19]

Heartfelt support for the Protocol's core-outcome rule among Annex I parties was always much less than the number of ratifications of the Protocol might suggest. Well before countries began to drop out in 2011, Annex I parties were

---

16 René Lefeber (2012), 'Climate Change and State Responsibility', in *International Law in the Era of Climate Change*, ed. Rosemary Rayfuse and Shirley V. Scott, p. 322.

17 Abram Chayes, Antonia Handler Chayes, and Ronald B. Mitchell (1998), 'Managing Compliance', p. 40.

18 In 2011, Japan and Russia refused to join the Kyoto Protocol's second commitment period: Kyoto Protocol (2011), *Decision 1/CMP.7, Outcome of the Work of the Ad Hoc Working Group on Further Commitments for Annex I Parties under the Kyoto Protocol at Its Sixteenth Session*, FCCC/KP/CMP/2011/10/Add.1, p. 5. In 2012, New Zealand announced that it would not sign up either: Kyoto Protocol (2012), *Decision 1/CMP.8, Amendment to the Kyoto Protocol Pursuant to Its Article 3, Paragraph 9 (Doha Amendment)*, FCCC/KP/CMP/2012/13/Add.1, p. 8. Canada completely withdrew from the Protocol in December 2012: see <http://unfccc.int/kyoto_protocol/background/items/6603.php>.

19 For the acceptance count, see <http://unfccc.int/kyoto_protocol/doha_amendment/items/7362.php>. For the amendment to enter into force, 144 state instruments of acceptance must be deposited.

pressing for the rule (compulsory mitigation) to be extended to developing countries. That demand, which was gathering force already in 2009, elicited fierce resistance from developing countries, which claimed that the responsibility for mitigation applied to the Annex I group alone. Under those conditions, the Protocol's rule could not lead to the further development of the FCCC's foundational law.

In 2010, Brunnée and Toope wrote about the international climate change regime thus:

> As far as specific substantive commitments are concerned, shared understandings do not currently extend much beyond the proposition that the existing regime be developed to include deeper commitments for more parties, with industrialized countries taking the lead. [R]obust agreement on interim targets for industrialized countries and long-term targets for major developing countries remains elusive.[20]

Nothing had changed by 2014 to diminish the accuracy of this comment. The 'pledge' period of mitigation action under the FCCC which began in 2013 (see 'B' and 'E' in Table 1.1) is a period of voluntary action that has not created a new core-outcome rule.

From this brief review of international climate change law it will be gleaned that the regime does not have much to offer, in a substantive way. The low profile of law in the IPCC reports is partially accounted for by this observation. (On the procedural level the regime has more to claim credit for, as I explain in Chapter 2.) At the same time, the meagre climate law that does exist at the international level enjoys a strong normative foundation; the community of states agrees that unchecked climate change is the greatest present threat to human well-being, and to much else besides.

A rounded account of the state of the law would also need to consider climate change law at the domestic level. International law is often informed by law that comes into being first at the level of states. It is sometimes possible to conclude from strong and consistent state legislative practice that a universal norm has 'crystallized', which thereafter is used to explain the observed uninterrupted uniformity in state practice. A brief review of the domestic scene is thus warranted.

1.3 Domestic climate law

At the domestic level, there are two possible sources of climate change law: statutory law and case law. In some countries, greenhouse gas emissions have been classed together with atmospheric pollutants. In such cases, large industrial emitters have incurred costs, through a tax or permit system, for releasing greenhouse gases into the atmosphere. Emissions have also been regulated through

---

20 Jutta Brunnée and Stephen J. Toope (2010), *Legitimacy and Legality in International Law: An Interactional Account*, p. 142.

emission standards (e.g. standards for motor-vehicle emissions). The regulatory costs of such measures are spread throughout society and impact on consumer prices. The classification of greenhouse gases as pollutants is a substantive legal reform, since it provides a basis for subsequent emission-reduction regulation in the state.

However, the engagement of pollution law as a means to curb emissions is not yet a practice so common, or so widely supported, or so frequently followed through with implementing regulation, as to suggest a general rule of law at work. No norm has yet evolved to the effect that carbon dioxide is a harmful pollutant that must be controlled. In the United States, the engagement of federal pollution law attests not to the success of climate change law in that country but to the total defeat of climate change bills in Congress. The US government is supportive of the FCCC and its foundational mitigation law, but the country is too divided on this issue to be able to develop a system of domestic climate change law. The best that the Obama Administration can do is to commandeer a law developed for another purpose, through which it might be able to deliver some mitigation benefits until such time as the political outlook improves.[21]

The 'operating environment' for climate law is difficult not only in the United States. Nevertheless, if one begins as I have by considering the impasse climate law has reached in that country, one is better equipped to recognize that the law's progress has been modest everywhere.

Even in jurisdictions where greenhouse gas emissions have been priced, and therefore in a sense penalized, emission-generating activities are in general not stigmatized as socially harmful. Normativity cannot easily gain a foothold where the problem is an *excess* in a certain kind of activity. Only a subset of anthropogenic emissions is ever penalized in the few countries that have such a system.[22] This is almost too obvious to state, yet, with the possible exception of deforestation, the activities that generate greenhouse gas emissions are still highly valued, both as investments and for the social goods they produce. Fossil-fuel infrastructure, some of it unimaginably vast (consider the Keystone XL Pipeline in the United States, the South Stream gas pipeline from Russia to Europe, or the recently approved Galilee Basin coal mine in Queensland, Australia),[23] is increasing the flow and reliability of supply of carbon-based

---

21 The US Environmental Protection Agency has found that greenhouse gases endanger public health and the environment and has listed them as pollutants, thereby triggering federal regulatory obligations. Implementing EPA rules (for the most recent of which see US Environmental Protection Agency (2014), *Carbon Pollution Emission Guidelines for Existing Stationary Sources: Electric Utility Generating Units: Proposed Rule*) is open to legal challenge and could introduce delays lasting years.

22 Deforestation has been the only emission-generating activity that has been successfully stigmatized as an idea in the mind of many, although that process began long before it was known that deforestation contributes to climate change.

23 And there is much more in the pipeline, so to speak. See Sierra Club Foundation (2012), *Annual Report 2012*, p. 13 ('Even if all proposed pipelines were built—TransCanada's Keystone XL among them—there still wouldn't be enough capacity to handle the growth that companies have laid out in their expansion plans').

energy. Prices for both coal and gas have remained low compared with other energy sources.²⁴ Natural-resource exploitation laws in domestic legal systems, but also philosophies such as libertarianism, have facilitated these developments. A greater number of legal principles support the received patterns of energy-use than oppose them. It is not surprising, in a fossil-fuel world, to find that domestic legal systems are stacked in favour of fossil fuels.²⁵ A growing middle class in all countries is consuming more, travelling more, living more energetically. In developing countries, fast-expanding utility grids are bringing fossil-fuel power to millions of new consumers.²⁶ Coal, in fact, will power much of the new demand.²⁷

These facts weigh down on climate law and severely limit its development. The machinery of the law is not stirred into action by intangible environmental problems caused by goods, or the modalities of making goods, that everyone uses and depends upon. Nor could normative law easily develop against a way of life that the great majority of people approve of and enjoy.²⁸ It is thus possible to deduce in an almost a priori manner that domestic laws that limit emission-generating activities within a state will be few and far between, and because

---

24 As the BBC reports, 'Gas and oil discoveries in shale rock in the US have led to a boom in gas and oil production there in recent years, and have also dramatically reduced gas prices.' It quotes the International Energy Agency as saying that 'the US will overtake Russia as the world's biggest gas producer by 2015, and Saudi Arabia as the world's biggest oil producer by about 2020.' John Moylan (2013), 'UK Shale Gas Reserves May be Bigger Than First Thought' <http://www.bbc.co.uk/news/business-22748915>.

25 For example, Germany's federal environment agency calculated that environmentally harmful subsidies in Germany through direct payments or tax benefits amounted to €42 billion per annum: Felix Ekardt (2013), 'Climate Law in Germany', in *Climate Change and the Law*, ed. Erkki J. Hollo, Kati Kulovesi, and Michael Mehling, p. 530. The IPCC's fifth assessment report states that governments are still spending far more money to subsidize fossil fuels than to accelerate the shift to cleaner energy: IPCC (2014), *5AR WG3 (Final Draft)*, chapters 7 and 13–15. See also IEA (2013), *World Energy Outlook 2013*, p. 81 ('Fossil-fuel subsidies increased to $544 billion per annum in 2012, more than five times the level of support for renewables').

26 Nearly 1.3 billion people did not have access to electricity in 2011: IEA (2013), *World Energy Outlook 2013*, p. 55.

27 Michael Kidd and Ed Couzens (2013), 'Climate Change Responses in South Africa', in *Climate Change and the Law*, ed. Erkki J. Hollo, Kati Kulovesi, and Michael Mehling, p. 637 ('untapped coal reserves in South Africa were estimated at 55 billion tonnes, and coal [will] remain the primary energy source [for the country] into the future'); and Intergovernmental Panel on Climate Change (2014), *Climate Change 2014: Mitigation of Climate Change: Working Group III Contribution to the Fifth Assessment Report of the Intergovernmental Panel on Climate Change: Summary for Policymakers*, p. 8 ('Increased use of coal relative to other energy sources has reversed the long-standing trend of gradual decarbonization of the world's energy supply').

28 In the United States, the activities of individual persons account for about one-third of CO_2 emissions, more than the entire US industrial sector: Michael P. Vandenbergh and Anne C. Steinemann (2007), 'The Carbon-Neutral Individual', 82 *New York University Law Review* 1673, p. 1694.

they will seem arbitrary as to the setting of the threshold amount above which emissions are penalized, people affected by them will have no straightforward way to locate their normative quality. This deduction is powered by both fact and logic: the anaemic condition of international climate law means that there is little pressure on domestic jurisdictions from above, while pressure from below can build only very slowly for a problem that is more global than local.

GLOBE's periodic surveys of domestic jurisdictions would seem, however, to suggest otherwise. The NGO, which rallies parliamentarians to act against climate change, claims to find 'substantive legislative progress' in the development of climate change legislation worldwide.[29] Yet its argument is tendentious: it creates momentum to work up further momentum because that is its fighting aim. GLOBE's surveys define 'law' loosely to include plans, programmes, and other varieties of non-legislated, executive policy, which are of course not law.[30] The NGO does not set out to distinguish action from the appearance of action. Most of the 'progress' it lists is unrelated to legal obligation.

The leader in climate change legislation since 2003 has been the European Union, a non-domestic, multinational jurisdiction. The main instrument for the mitigation of emissions in the EU is Directive 2003/87, establishing a scheme for trade in emission allowances. In addition, there is the Climate and Energy Package 2008, which comprises four pieces of legislation, including the so-called Revision and Strengthening of the EU Emissions Trading Scheme and Reducing GHG Emissions Fairly. Another element of EU climate law is Directive 2009/28, whose aim is to increase renewable-energy generation in the EU to 20 per cent of energy by 2020. The EU also has a non-binding target for increasing energy efficiency; specific laws aim to assist its achievement, such as a 2002 directive on the energy efficiency of buildings. In February 2014, the European Parliament endorsed a new target for CO_2 emissions from cars.[31] Of these measures, some have the potential to reduce emissions in certain sectors of the EU economy below historical levels, while others aim to moderate the rate of emission growth. Concurrently, laws of EU member states provide support for fossil fuels (coal is an important source of energy in Poland, Spain, and Germany). EU policy (but not law) seeks to phase out state aid for coal by 2018.[32] The EU's climate change legislation is neither

---

29 Michal Nachmany et al. (2014), *The GLOBE Climate Legislation Study: A Review of Climate Change Legislation in 66 Countries (4th edn)*, p. 4.
30 Ibid., p. 37. The study defines climate change law as 'Legislation, or regulations, policies and decrees with a comparable status, that refer specifically to climate change or that relate to reducing energy demand, promoting low carbon energy supply, tackling deforestation, promoting sustainable land use, sustainable transportation, or adaptation to climate impacts' (ibid.).
31 European Commission (2014), 'Commissioner Hedegaard Welcomes Agreement on the Car Emissions Target', <http://europa.eu/rapid/press-release_STATEMENT-14-25_en.htm>.
32 Ludwig Krämer (2012), 'European Union Law', in *Climate Change Liability: Transnational Law and Practice*, ed. Richard Lord et al., p. 360.

comprehensive[33] nor closely enforced.[34] The main reason for the attention it enjoys is that so few countries have followed the EU's example or attempted to go as far as it has. Certainly, the EU has enacted some 'core-outcome' climate law. Its emission-trading scheme has priced emissions and forced companies to consider mitigation measures that they would not otherwise have considered. The EU's initiatives on climate legislation may have helped EU member states to meet their targets under the Protocol's first commitment period.[35] The Global Financial Crisis which hit Europe in 2007–2008 reduced economic activity and emissions for years thereafter, which meant that a high carbon price was avoided. Carbon prices collapsed more than once. In brief, the EU case offers some support for the thesis that a climate law has become established in EU member states. But the system has not yet operated with a high pressure on member states.

A country with relatively strong domestic climate laws is the United Kingdom. Reid describes the Climate Change Acts passed by the UK parliament in 2008 as 'innovative in establishing legally binding targets for emissions reductions'.[36] The law imposes a legal duty on government ministers to achieve the reductions. (The target for 2020 is a 34 per cent reduction from 1990 levels.) Policy and regulatory initiatives to support the targets include incentives for improved home insulation, higher building standards, and a 'renewables obligation' on electricity generators.[37] Regular reporting to parliament on progress with meeting five-yearly 'carbon budgets' is compulsory. Reid notes that the legislation provides no sanctions for a failure to meet the targets, and the precise legal status of the ministerial duty is unclear: 'there are arguments that these provisions should be interpreted in line with previous target-setting legislation which merely requires Ministers to make reasonable endeavours to see that they are achieved'.[38] He concludes that the effectiveness of the legislation in achieving significant emission reductions will depend on political and public pressure rather than on the law itself.[39]

The UK is, of course, an EU member state. Its domestic system will facilitate the country's compliance with the burden it has accepted under the EU's mitigation target. The conclusion about the UK is the same as that about the EU: the evidence suggests a remarkable normative development in favour of mitigation (politically consensual, socially supported, and not likely to be undone), but as yet one of modest ambition, possibly aiming no higher than the facilitation of the UK Kyoto Protocol target for the second commitment period.

---

33 Ibid., p. 353.
34 The European Commission 'has not taken action against member states when the climate change provisions of EU law had been incorrectly applied, for example when greenhouse gas emissions had increased': ibid., p. 366.
35 See Section 4.2.2 in Chapter 4.
36 Colin T. Reid (2013), 'Climate Law in the United Kingdom', in *Climate Change and the Law*, ed. Erkki J. Hollo, Kati Kulovesi, and Michael Mehling, pp. 537–538.
37 Ibid., pp. 539–541.
38 Ibid., p. 542. See also Silke Goldberg and Richard Lord (2012), 'England', in *Climate Change Liability: Transnational Law and Practice*, ed. Richard Lord et al., p. 447 ('the government could be liable for judicial review'). No such action has yet materialized.
39 Colin T. Reid (2013), 'Climate Law in the United Kingdom', pp. 546 and 549.

Core-outcome climate law that limits country emissions by reference to emissions in a base year may be briefly summarized for other domestic jurisdictions.[40] France's 'Grenelle' laws (Grenelle I and II, 2009–2010) include measures on emission targets, as well as on renewable energy and energy efficiency.[41] Switzerland's Revised CO_2 Act 2013 sets emission-reduction targets of 20 per cent below 1990 levels by 2020 (identical to the EU) as well as interim goals for buildings, transportation, and industry.[42] Looking outside of Europe for examples, New Zealand is a country with ostensibly advanced climate laws aimed at core mitigation outcomes. However, New Zealand has refused to take on legally binding targets for the Protocol's second commitment period. New Zealand's Climate Change Response Act 2002 established a legal framework to facilitate the country's compliance with its obligations under the international climate change regime. The country has an emission-trading scheme that operates without a formal cap on emissions. A review by an independent panel appointed by the government in 2011 concluded that the NZ ETS was not incentivizing low-carbon investment and had not led to significant reductions in emissions.[43] In August 2013, New Zealand announced that its FCCC mitigation target for 2020 would be a 5 per cent reduction from 1990 emission levels.[44] (Its Kyoto Protocol target for 2008–2012 was to return to 1990 levels.) Mundaca and Richter have criticized New Zealand's climate laws as incomplete and inadequate. They argue that an ambitious programme of caps is needed to increase investment in clean energy in the country.[45] For the moment, New Zealand has an engine of change in place but chooses to keep it idle.

Japan's 1998 Law Concerning the Promotion of Measures to Cope with Global Warming, amended in 2005, was the country's implementing instrument for its so-called Kyoto Achievement Plan. The targets set were voluntary for domestic actors. According to Kimura, Japan's emission-trading scheme 'attracts participants that can easily achieve the pledged targets'.[46] In 2005, Japan was highly dependent on nuclear energy (accounting for 24 per cent of total electricity generation) and had plans to increase its proportion to 30 or 40 per cent of the energy mix.[47] Its climate laws thus rested on the assumption that nuclear energy would continue to be plentifully available. In 2011, following the Fukushima accident, Japan decided that it would phase out nuclear energy and aim to secure 20 per cent of power generation from renewables.[48] In a further policy change in 2013, the country abandoned its FCCC mitigation target of 25 per cent below

---

40 See Michal Nachmany et al. (2014), *GLOBE Climate Legislation Study*, pp. 9–22.
41 Ibid., p. 203.
42 Ibid., p. 535.
43 Luis Mundaca and Jessika Luth Richter (2014), 'Challenges for New Zealand's Carbon Market', 3 *Nature Climate Change* 1006, p. 1007.
44 Ibid., p. 1006.
45 Ibid., p. 1007.
46 Hitomi Kimura (2013), 'Climate Law and Policy in Japan', in *Climate Change and the Law*, ed. Erkki J. Hollo, Kati Kulovesi, and Michael Mehling, p. 588.
47 Ibid., p. 591.
48 Ibid., p. 592.

2005 levels by 2020. Its current aim is to cut emissions by just 3.8 per cent from 2005 (equivalent to a 3.1 per cent *increase* over 1990 levels). Because Japan is not participating in the Protocol's second commitment period, its current target, under the FCCC, is not legally binding. The country has further announced that it cannot commit to the renewable energy target it adopted in 2011.[49]

I return briefly to the situation in the United States. As is well known, the country has never been a party to the Kyoto Protocol. This has been a Democratic as much as a Republican stance. The domestic political consensus is that the country should not consider agreeing to ambitious emission limits unless high-emitting developing countries also accept limits. A well-founded domestic climate law could hardly arise in this situation of an across-the-spectrum rejection of the presumptions of the current international regime. And indeed, as Hilson observes, 'in formal terms one might struggle to call the US federal regime one involving climate law at all'.[50] The Obama Administration has sought, as we noted, to moderate emissions from the transport and energy sectors by enlisting the help of the Clean Air Act—a second-best solution,[51] because it is a stop-gap measure that does not create a regime unique to climate change. Instead, it somewhat implausibly expands existing environmental laws.[52] Canada has taken a similar approach.[53] There are many reasons to promote domestic mitigation outside of legal obligation (economic efficiency and technological advantage are among them), and both the United States and Canada at the federal level have lent it support, up to a point. Yet there is no evidence of a relevant normative development in the legislative activities of

---

49 *The Japan Times*, 19 November 2013, <http://www.japantimes.co.jp/opinion/2013/11/19/editorials/cut-emissions-without-nuclear-power/>.
50 Chris Hilson (2013), 'It's All About Climate Change, Stupid! Exploring the Relationship between Environmental Law and Climate Law', 25 (3) *Journal of Environmental Law* 359, pp. 366–367.
51 As it has been called by John Copeland Nagle (2010), 'Climate Exceptionalism', 40 *Environmental Law* 53, p. 73. On the United States' various uses of the Clean Air Act, see Michael B. Gerrard and Gregory E. Wannier (2012), 'United States of America', in *Climate Change Liability: Transnational Law and Practice*, ed. Richard Lord et al., p. 565, and US EPA (2014), *Carbon Pollution Emission Guidelines*.
52 For a pollutant to be subject to Clean Air Act requirements, it must be deemed by the EPA to endanger public health and welfare. Carbon dioxide has this effect above a certain atmospheric concentration, but the concentration buildup is largely determined by countries other than the United States. US territorial emissions would not endanger public health and welfare were it not for those other emissions. See Eric A. Posner and David Weisbach (2010), *Climate Change Justice*, p. 70 ('Existing pollution statutes were not designed with climate change in mind, and thus did not give the EPA the authority that it would need to implement a sensible regulatory system').
53 The federal Canadian Environmental Protection Act was amended to add six greenhouse gases to the 'List of Toxic Substances' in Schedule 1 of the act; Jane Matthews Glenn and José Otero (2013), 'Canada and the Kyoto Protocol: An Aesop Fable', in *Climate Change and the Law*, ed. Erkki J. Hollo, Kati Kulovesi, and Michael Mehling, p. 497.

either country. In Canada's case, core-outcome measures put in place by one government were thrown out by the next.[54]

Recent events in Australia confirm that climate law does not have a secure foundation in domestic legislation. A comprehensive mitigation law promulgated in 2011 by an Australian centre-left government to set up a mechanism to control the country's greenhouse gas emissions through a polluter-pays system was undone in 2014 by a conservative government elected on a platform of non-prescriptive action on climate change. The 2011 law provided for the pricing of carbon emissions in key sectors of the Australian economy. In substance, it said: 'Greenhouse gases are to be deemed a form of pollution for which the government will collect a fee.'[55] Under the scheme, consequential higher prices for emission-intensive consumables, especially electricity, as well as a new stream of income from the sale of emission allowances for the government to spend on support for renewable energy and energy efficiency (among other mitigation measures) would reduce Australia's net emissions to around the 1990 level for the first time. Even with the new law in place, the country's gross emissions were forecast to grow for decades to come, because the Labor government that passed it had decided that it would operate the lever of the pricing mechanism with a light touch and would buy emission allowances on the international market to achieve the net reduction. In particular, the capping of emissions inside the country was to be no more onerous than limits set by other developed countries.

Australia's imposition of a fee on carbon emissions, enshrined in national legislation, was a direct attack on the causes of climate change. In a global context, even in 2011, it was an uncommon measure. By raising the prices of high-emission goods, the government would be taking money away from everyone. Much of it would be returned to the least well off to assist them with the price rises. The fact that all citizens would be made to contribute to the cost of cutting emissions—and thus that it was not a cost-neutral measure—is itself a sign of the measure's genuineness.[56] The Labor government sought to represent

---

54 Ibid., p. 489 ('in 2006 ... the pro-Kyoto government was replaced in Ottawa by a right-of-centre anti-Kyoto one, which set about dismantling previous climate change initiatives and replacing them with their own, less stringent ones'). On the general point about normative development, see Eric A. Posner and David Weisbach (2010), *Climate Change Justice*, p. 59 ('Most of the legislation and other forms of government action that have taken place [worldwide] have been largely symbolic'); and Michael Mehling (2013), 'Implementing Climate Governance: Instrument Choice and Interaction', in *Climate Change and the Law*, ed. Erkki J. Hollo, Kati Kulovesi, and Michael Mehling, p. 11 ('the countless rules devoted to climate change are still but loosely related and far from becoming a coherent normative framework').

55 For the logic of the act, see Clean Energy Act 2011 (repealed), section 4.

56 Climate mitigation is costly. Climate change 'requires policies that will affect nearly everyone—often hitting individuals in the pocket through higher prices for energy': Paul G. Harris (2014), 'Risk-Averse Governments', 4 *Nature Climate Change* 245, p. 246. According to the IPCC, if all countries in the world were to begin mitigation immediately, if there were a single global carbon price, and if all key technologies were available, mitigation scenarios that stabilize atmospheric concentrations of CO_2 eq. at around 450 ppm by 2100 entail losses in global consumption of 1–4% in 2030,

the core rule of its 2011 law as an act of compliance with international obligations ('The objects of this Act are ... to give effect to Australia's obligations under ... the Climate Change Convention [and] the Kyoto Protocol').[57] In 2013, the centre-right of federal politics took over government. It immediately prepared legislation to repeal the 2011 law, and the country began to revert to a system of indirect, voluntary, and incentive-based mitigation measures. The repeal was effected in July 2014. Along with carbon pricing, Australia's new government jettisoned the country's legislated support for (in the words of the original act) an 'effective global response to climate change, consistent with Australia's national interest in ensuring that average global temperatures increase by not more than 2 degrees Celsius above pre-industrial levels'. It also removed the legislated long-term target 'of reducing Australia's net greenhouse gas emissions to 80 per cent below 2000 levels by 2050'.[58] Having backflipped on this rare instance of climate law, Australia was not called upon to account for its actions before any international forum. No other state claimed, for example, that Australia had ceased to be compliant with its international legal obligations. In reality, Australia simply rejoined the mainstream of states that have no substantial climate laws.

The Australian reversal is even more remarkable when the country's physical circumstances are considered. Australians have experienced a changing climate as intensely as any other people. In 2013, when the new government began to dismantle the country's climate legislation, Australians had lived through extreme warmth. January 2013 was the hottest month ever recorded in the country.[59] Monthly maximum temperature anomalies reached +5 °C in parts of the state of New South Wales. Widespread heat across Australia contributed to this record, rather than extreme heat at individual locations. The extreme warmth gave the country its hottest summer (December 2012 – February 2013) on record. This was followed by the third warmest winter (June–August 2013) on record. Temperatures for September 2013 were 2.75 °C above average, the highest monthly temperature departure ever recorded in Australia. The all-time warmest 12-month period in Australia was observed between September 2012 and August 2013, but the record was broken again over the next two months, making November 2012 to October 2013 the new warmest 12-month period on record. Voters were well aware of the temperature trends, yet they elected government opposed to directive climate change legislation. Even fully backed by an unfolding of extreme weather events

---

2–6% in 2050, and 3–11% in 2100 relative to consumption in baseline scenarios that grow anywhere from 300% to more than 900% over the century. 'These numbers correspond to an annualized reduction of consumption growth by 0.04 to 0.14 (median: 0.06) percentage points over the century relative to annualized consumption growth in the baseline that is between 1.6% and 3% per year.' IPCC (2014), *5AR WG3 SPM*, p. 15.

57 Clean Energy Act 2011 (repealed), s. 3.
58 Clean Energy Act 2011 (repealed), s. 3. The 2011 law did not inscribe a short-term target.
59 For this and the following information, see World Meteorological Organization, *Provisional Statement on Status of the Climate in 2013*, available from <www.wmo.int/pages/mediacentre/press_releases/pr_981_en.html>.

in real time, mitigation law has had no normative purchase on the Australian body politic.

In a domestic context, climate change law could also be developed in the courts. Given the paucity of international treaty law and national legislation as founts of climate law, it is unlikely as an a priori matter that courts in any country will have been able to develop a law on the mitigation of emissions. It might be just possible to successfully sue on traditional legal grounds (property damage, other economic loss, nuisance, negligence, etc.) for compensation for, or protection from, climate change *impacts*,[60] but this would not count as climate law according to my definition of what a cohesive climate law would encompass. This type of action, if available, would merely be tort or planning law applied to new facts allegedly caused by climate change in the past or expected in the future.[61] Much climate litigation has been purely political, in the sense that it has had no plausible legal basis or chance of success. Its objective has been to shame governments for their inaction, but it is entirely unable to build a mitigation law.[62] What it comes down to is climate politics by other means. It almost always fails to gain traction, because the government targeted by the action invariably has not breached any law.

---

60 No action of this type has yet succeeded: Jutta Brunnée et al. (2012), 'Overview of Legal Issues Relevant to Climate Change', in *Climate Change Liability: Transnational Law and Practice*, ed. Richard Lord et al., pp. 23, 27, 32–33. See also Giedrė Kaminskaitė-Salters (2011), 'Climate Change Litigation in the UK: Its Feasibility and Prospects', in *Climate Change Liability*, ed. Michael Faure and Marjan Peeters, p. 165 (private litigation for climate damage is likely to gain in prominence *in the future*, speculates the author, but is non-existent in the present). Public-nuisance cases in the United States have become 'the largest source of climate-relevant private litigation today', write Michael B. Gerrard and Gregory E. Wannier (2012), 'USA', pp. 581–583. These cases have so far all been dismissed without success for the plaintiffs.

61 Jutta Brunnée et al. (2012), 'Overview', pp. 29–33 ('Climate change liability may be engaged in relation to laws or regulations that do not have climate change control as their primary object. [A]dministrative decisions on planning and permits for projects of many kinds, from mines to dams to power plants, may be subject to litigation on the grounds that climate change considerations have not been taken into account'). These comments are about what is possible in principle rather than anything that happens in practice.

62 Nevertheless, it is praised! Support for climate litigation of the political kind is expressed, for example, by David B. Hunter (2009), 'The Implications of Climate Change Litigation: Litigation for International Environmental Law-Making', in *Adjudicating Climate Change: State, National, and International Approaches*, ed. William C. G. Burns and Hari M. Osofsky, pp. 358–360; and Giedrė Kaminskaitė-Salters (2011), 'Climate Change Litigation in the UK', p. 170. See also Eric A. Posner and David Weisbach (2010), *Climate Change Justice*, p. 71; and Chris van Dijk (2011), 'Civil Liability for Global Warming in the Netherlands', in *Climate Change Liability*, ed. Michael Faure and Marjan Peeters, p. 208. For an account of politically driven climate litigation in Canada, see Jane Matthews Glenn and José Otero (2013), 'Canada and the Kyoto Protocol', p. 499 (on the *Friends of the Earth v. Canada* case); and Silke Goldberg and Richard Lord (2012), 'England', p. 456 (on the UK's 'hopeless' *RBS* case).

A possible exception to the general impotence of mitigation litigation is the case of *Massachusetts v. EPA*, which resulted in the US EPA branding greenhouse gases as pollutants under US law, making them the subject of further government regulation, and thus potentially affecting emission standards for US motor vehicles and power plants.[63] In general, however, domestic climate litigation has been irrelevant to the development of a climate change law, and in most jurisdictions it has barely registered at all.[64]

1.4 A perspective for law

Despite the poverty of climate law as I have described it, there is much that can be said about state compliance with the few rules that do exist. The notion of compliance has been variously defined in this context.[65] The definition I employ

---

63 For a discussion, see Michael B. Gerrard and Gregory E. Wannier (2012), 'USA', p. 565. The United States is described as having experienced a large surge in litigation activity, 'from only one climate-related case brought in 2003, to over a hundred cases in 2010'. An increase in climate-related litigation is not necessarily evidence of a climate law being developed in the courts. The US cases were brought either by environmental groups seeking to pressure government to introduce climate change regulation, or by industry or states (more than 90 of the cases mentioned) opposing the regulatory innovations based on the Clean Air Act: ibid., pp. 562, 565–566.

64 See Giedrė Kaminskaitė-Salters (2011), 'Climate Change Litigation in the UK', p. 171 ('no cases testing the feasibility of climate-based litigation have been brought to the English Courts to date'); Chris van Dijk (2011), 'Civil Liability', p. 208 ('the chance of a claim being allowed—in any case in the Netherlands—is definitely small at the moment', and no such claim had been allowed at the time of writing); Ross Abbs, Peter Cashman, and Tim Stephens (2012), 'Australia', in *Climate Change Liability: Transnational Law and Practice*, ed. Richard Lord et al., pp. 76–81 ('all mitigation cases failed in their primary objective, and at most succeeded in obliging the decision-maker to "take account" of greenhouse gas emissions from the proposed development'); Meinhard Doelle, Dennis Mahony, and Alex Smith (2012), 'Canada', in *Climate Change Liability: Transnational Law and Practice*, ed. Richard Lord et al., p. 555 ('Climate change litigation in Canada is in its infancy'); Silke Goldberg and Richard Lord (2012), 'England', pp. 487–488 (reiterating the point that 'in the UK, no specific regime for climate change liability has developed as yet and the jurisprudence in this area is still in an early phase'); Ludwig Krämer (2012), 'EU Law', pp. 374–375 ('Litigation initiated at EU level in favour of more drastic adaptation or mitigation measures is [non-existent to date and is] unlikely to increase in the future'); and Jolene Lin (2014), 'Litigating Climate Change in Asia', 4 (1) *Climate Law* 140 (non-existent).

65 See, e.g., Kal Raustiala and Anne-Marie Slaughter (2002), 'International Law, International Relations and Compliance', p. 539 ('a state of conformity or identity between an actor's behavior and a specified rule'); Gregory Rose and Lal Kurukulasuriya (2007), *Compliance Mechanisms under Selected Multilateral Environmental Agreements*, p. 19 ('the fulfilment by the contracting parties of their obligations under a multilateral environmental agreement'); Michael G. Faure and Jürgen Lefevere (2011), 'Compliance with Global Environmental Policy', in *The Global Environment: Institutions, Law, and Policy*, ed. Regina S. Axelrod, Stacy D. VanDeveer, and David Leonard Downie, p. 173 ('the extent to which the behaviour of a state party to an international treaty actually conforms to the conditions set out in that

in this book is as follows: state compliance is the implementation by a state of the actions it is obliged to implement in accordance with a treaty regime to which it is party. 'Non-compliance', by implication, is conduct that deviates from treaty rules.

The fact that the causes of climate change subsist the world over in the lawful conduct of citizens and states must account for much of the difficulty that academic lawyers have in making their own discipline relevant to the problem of climate change. The legal system has been friendly to fossil fuels and has not developed fundamental principles or rules of such force and general acceptance as to make a positive contribution to the control of climate change. No wonder the IPCC has almost nothing to say about climate change *law*.

In situations where conventional sources of law contribute little to constrain the identified social/physical threat and where the relevant actions are rarely or only lightly regulated and over-emitters can always buy their way out of trouble (e.g. through the acquisition of emission allowances created by offset projects), law scholars can and do fall back on regulations, policies, and informal arrangements, bearing on, for example, indirect mitigation activities, such as incentives for energy efficiency, renewable-energy targets, or the implementation of emission trading or offset markets. These measures generally respond to greenhouse-gas production indirectly, not by laying down a rule that restricts emission of the gases, but by reducing waste or supporting the development of alternatives to fossil fuels, thereby allowing for greenhouse gas emissions to continue, and indeed to increase, although possibly at a slower rate than would have been the case without the measures. Legal scholarship that focuses on such peripheral controls ends up being rather diffuse, and not always evidently about climate change per se; for the scholarship tends to blend in with such pre-existing specialty areas as property rights, consumer-protection law, constitutional or administrative law, the estimation of the costs of alternative policies and their distributional benefits, and the assessment of instrument choice and design from the perspective of fairness or effectiveness.

While work along these lines develops our understanding of traditional law topics, it is unable to represent legal expertise as being part of the *solution* to the climate change threat. It has the flavour of a parallel (academic, technical, derivative) conversation. What else is there, then, for academic lawyers to contribute to a cross-disciplinary effort to control climate change?

One cure for law's marginalization is for legal scholars to seek to throw light on the very nature of climate change control emanating from the level of the

---

treaty'); and Jørgen Wettestad (2007), 'Monitoring and Verification', in *The Oxford Handbook of International Environmental Law*, ed. Daniel Bodansky, Jutta Brunnée, and Ellen Hey, p. 975 ('the relationship between state actions and international commitments—and responses to possible mismatches'). Mitchell has distinguished 'compliance', meaning an actor's behaviour that conforms to a treaty's explicit rules, from 'treaty-induced compliance', which is behaviour that occurs because of a treaty's compliance system: Ronald B. Mitchell (1993), 'Compliance Theory: A Synthesis', 2 (4) *Review of European Community and International Environmental Law* 327, p. 328.

FCCC and the implementation systems it has given rise to. Academic lawyers are best qualified to detect the emergence of a law, and to distinguish it from that which purports or is merely assumed to be a law. An excellent example of this kind of analysis is an account of South African climate change law by Kidd and Couzens, which brings out domestic contradictions, window-dressing, and government cynicism about climate change. Their expertise in law enables them to see past the official veneer and highlight an actual condition of inaction.[66]

Much has been written on the question 'Why do states obey international law?' This book asks not why states obey climate change law but whether the climate change regime has generated any law, and, if so, whether states have complied with it. What control has been achieved, really, how does it work in practice as opposed to theory, what is the difference between what it purports to be doing and what it actually delivers, how far does the control extend, where does it break down, and where is it illusory? If law of some kind is finally needed to control climate change, then a service that law academics can perform is to clarify the current legal landscape for what it is.

The focus on 'compliance' in international climate change law revolves clearly enough around a fundamental legal concern (the question of state subjection to law). It is itself an entryway into the interrogation of the nature of the rules and legal obligation that constitute the climate change regime. In undertaking this investigation, I use the international regime as my anchor. I have limited the scope of the book to the web of regulation created by the FCCC's and Kyoto Protocol's state parties. The FCCC, as elaborated through party decisions, purports to be rule-governed and to have created legal obligations in states to follow certain rules. The same is true of the Kyoto Protocol. In this approach, legal scholarship, by sweeping away myths and highlighting gaps, contributes to the critique of our efforts to control climate change.

---

66 Michael Kidd and Ed Couzens (2013), 'Climate Change Responses in South Africa', pp. 619, 634, and 638.

2 State compliance with rules on national reporting and assessment

2.1 Introduction: Importance of reporting

The international climate change regime's rules on state reporting of mitigation actions and results form a complex web. Rules on reporting under the FCCC and the Kyoto Protocol are legally binding on parties, mostly on Annex I parties, but some of these rules are binding on all parties. They are generally in the form of decisions of the Conference of the Parties elaborating the laconic directions in the treaty text of the FCCC and the Protocol.

State reporting necessitates monitoring of greenhouse gas emissions, mitigation actions, and other information at the state level. States must submit their reports to the FCCC Secretariat, which in turn makes them available to all regime parties and (with some exceptions) to the public. International assessment of the reported information is a separate exercise. Assessment engages the treaty parties in the scrutiny of the reports. Because the Protocol's Annex I parties can trade in emission allowances, state reporting and assessment must also take account of emission trading when it occurs. To track state trade in emission allowances, an international transaction mechanism has been set up.

This chapter reviews the rules on state reporting and assessment under the FCCC and the Protocol, the legal status of the rules, and the extent to which states have complied with their reporting and assessment obligations. In comparison with topics dealt with in subsequent chapters, the rules on state reporting are well developed and rarely the cause of conflict among states. This is especially true in the application of the rules to Annex I parties.

For almost two decades now, Annex I parties have been filing hefty 'national communications' with the FCCC Secretariat at regular intervals. Moreover, state greenhouse gas inventories with information going back to 1990 have, since around 2002, been submitted annually by Annex I parties. Non-Annex I parties have also been participating in reporting on emissions and on policies and measures on climate change, albeit more sporadically and less accurately and fully. A certain amount of reluctance to accept reporting responsibilities at the same level as Annex I parties, as well as a genuine difficulty in preparing complex and expensive reports at regular intervals, characterizes the position of many non-Annex I parties, yet all non-Annex I parties have filed reports at one time or another despite these issues.

By comparison, the international assessment of state reports is a more sensitive matter, even for Annex I parties. Nevertheless, as of 2010, some degree of international scrutiny of the reports of *all* countries has been accepted as necessary and is being put into practice.

A single imperative lies behind these aspects of 'accountable reporting': because states share responsibility for the global problem of climate change, they must be open about their contributions to it and about their actions to mitigate its causes.

The extent and regularity of state practice by both developed and developing countries in reporting information relevant to the climate change regime suggest that this is an area of the regime in which binding legal rules represent a consensus about the conduct necessary to meet the regime's objectives. The accountable reporting rule is generally carefully adhered to, unlike the general mitigation rule (Chapter 4) which in important respects is not complied with despite having an equal legal status to the reporting rule.

The evidence of state conduct reviewed in the following pages might suggest that the obligation of states to transparently report their emissions and actions on climate change mitigation is hardening into a rule of customary international law, one that binds states irrespective of any treaty, until such time as the danger of human-induced climate change has receded.[1] However, as will become clear in this and subsequent chapters, a customary-law argument, while plausible, cannot be sustained, due to the number of exceptions in state practice.

2.2 State obligations

2.2.1 Two treaties, two systems

Article 4 of the FCCC (on party 'commitments') is divided into obligations common to all parties, obligations pertaining only to Annex I parties, and obligations exclusive to Annex II parties (the old OECD group). The general commitments in article 4 relating to all parties are themselves internally differentiated: they neither impose equal obligations on all states nor demand undifferentiated adherence. They are, instead, subject to the parties' 'common but differentiated responsibilities and their specific national and regional development priorities, objectives and circumstances'.[2]

This so-called CBDR principle inevitably renders certain FCCC treaty obligations difficult to interpret because it relativizes them to country circumstances

---

1 Customary international law is typically defined as a 'general and consistent practice of states followed by them from a sense of legal obligation': Jack L. Goldsmith and Eric A. Posner (1999), 'A Theory of Customary International Law', 66 (4) *University of Chicago Law Review* 1113, p. 1113. The fact that little more than 20 years have elapsed since the FCCC was opened for signature is no objection, in theory, to an 'elevation' of any of its principles to CIL status.
2 FCCC, art. 4.

in ways that are not fully specified in the treaty text or in subsequent decisions of the parties.

Under article 12 of the FCCC, all state parties have reporting obligations. Reporting is to be channelled through national communications and emission inventories. A national communication is to provide information on a country's policies and measures in furtherance of various FCCC requirements (primarily relating to mitigation and adaptation), financial assistance and technology transfer to non-Annex I parties (if the national communication is that of an Annex I party), and assorted other actions (such as actions undertaken domestically by a country to raise public awareness about climate change). Moreover, a national communication is to provide information on a country's emissions[3] and projected emissions under different scenarios.

Greenhouse gas inventories, which Annex I parties must submit separately to national communications and with greater frequency, detail activity data (anthropogenic emission sources reported in standardized source types and quantified to show the amount of 'activity' under each source), emission factors (the emissions per unit of activity per source), and the methods used by the state's agency charged with the preparation of the national inventory to collect and synthesize the required information.

The Kyoto Protocol is, in all matters, narrower in scope than the FCCC. It is directed primarily at Annex I parties.[4] Article 7 of the Protocol requires developed-country parties to submit national communications and greenhouse gas inventories at regular intervals. Since these two outputs are already prepared and submitted under the FCCC, the Protocol's requirements can be satisfied by expanding the FCCC reports to include information addressing the additional requirements of the Protocol. For example, the Protocol enables emission allowances to be created from afforestation and reforestation activities (Emission Reduction Units), necessitating that these activities are accounted for separately from information required by the FCCC.

Under both treaties, as noted, state reports are subjected to some degree of international, expert-based scrutiny. To avoid duplication, the assessment procedures for the two treaties overlap significantly. In the Protocol's case, the assessment of Annex I reports is not only international and expert-based but also potentially interventionist, as the assessing body can take action where there is evidence of a party's non-compliance with the rules. The following four sections review the obligations of states on reporting and assessment. The second half of the chapter assesses state performance against these obligations. The Protocol's non-compliance system warrants treatment in a separate chapter (Chapter 3).

---

3 In the case of Annex I parties, this can be duplicative of the separately submitted emission inventories, although the latter cover this element of information in much greater detail than the national communications, and must be submitted annually.

4 The Clean Development Mechanism is the main exception, as it engages non-Annex I parties through its own system of rights and obligations.

2.2.2 National communications

The first set of rules to consider is the one applying under the FCCC system to Annex I parties. The term 'guideline', instead of 'rule', is the diplomatic choice of words used in party decisions for what I refer to here as rules. In practice the guidelines are treated as rules. Even the Protocol's non-compliance system is based on 'guidelines', as discussed in Chapter 3.

Guidelines for the preparation of national communications by Annex I parties were adopted in 1999.[5] Their aim was (and remains) to assist Annex I parties to submit 'consistent, transparent, comparable, accurate, and complete information' in order to enable a 'thorough review and assessment of the implementation of the Convention'.[6] What is not stated, but may be assumed to be an equally important aim of the guidelines, is that the sharing of information would build trust and enable a deeper cooperation among the parties. An annex to the 1999 decision provides a structure for national communications. To ensure their completeness, 'no mandatory element shall be excluded'.[7] As I explain below, it is not clear what the concept of 'mandatory element' refers to. Nevertheless, the point to emphasize here is that the 1999 decision elaborating the FCCC's reporting obligation treats the 'guidelines' as being, at least in the 'mandatory' part, legally binding.

For the purpose of Annex I national communications, a complete inventory of greenhouse gas emissions is not required, as it is provided separately.[8] Summary information on emissions for the period from 1990 onward is sufficient.[9]

In reporting on mitigation activities, Annex I parties are expected to quantify the expected impact of each of their policies and measures. No standardized methodologies exist for doing so.[10] Several elements to be included in national communications presuppose a sophisticated analysis as well as a degree of speculation, and this is particularly true of the assessment of future impacts. Parties must report on how their 'national circumstances' affect greenhouse gas emissions and removals, and how national circumstances and changes in those circumstances affect emissions and removals over time. National circumstances include the structure of government, population profile, geographic profile, climatic profile (temperature distribution, temperature variations, precipitation distribution, etc.), economic profile, energy resource base, prevailing transportation modes and

---

5 FCCC (1999), *Decision 4/CP.5, Guidelines for the Preparation of National Communications by Parties Included in Annex I to the Convention, Part II: UNFCCC Reporting Guidelines on National Communications*, FCCC/CP/1999/6/Add.1, para. 1. These guidelines are found in FCCC (1999), *Guidelines on Reporting and Review*, FCCC/CP/1999/7. Submission dates for Annex I national communications are determined by decisions of the Conference of the Parties; see e.g. FCCC (2007), *Decision 10/CP.13, Compilation and Synthesis of Fourth National Communications*, FCCC/CP/2007/6/Add.1.
6 FCCC (1999), *Guidelines on Reporting and Review*, section II, para. 1.b.
7 Ibid., section II, para. 5.
8 See Section 2.2.3.
9 FCCC (1999), *Guidelines on Reporting and Review*, section II, paras 10–11.
10 Taryn Fransen (2009), *Enhancing Today's MRV Framework to Meet Tomorrow's Needs: The Role of National Communications and Inventories*, p. 7.

32 *National reporting and assessment*

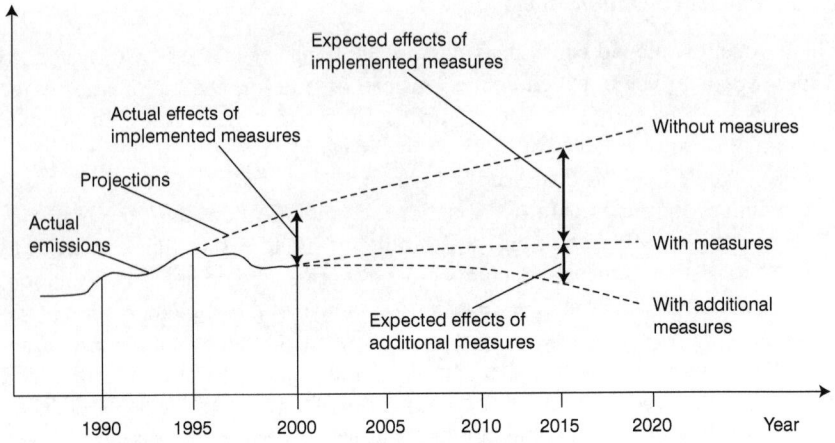

Figure 2.1 Varieties of emission projections required for an Annex I party's national communication. In this figure, the present is the year 2000. The solid curve represents actual emissions and the dotted curves represent projections. 'Additional measures' refers to mitigation measures that the country has not adopted but could adopt if it wished to (source: FCCC (1999), *Guidelines on Reporting and Review*, section II, para. 38).

travel distances, industrial profile, characteristics of building stock and urban structure, agricultural profile, and forest types and forest-management practices.[11]

The objective of the projections section of the national communication is to give an indication of future trends in emissions in the light of current national circumstances, including and excluding the government's adopted policies and measures, and including and excluding possible 'additional' measures. In other words, parties are required to provide an estimate of the total effect of their policies and measures in terms of emissions avoided compared with a situation without such policies and measures.[12] A graphic illustration of these rather speculative required elements of national communications is presented in Figure 2.1.

In interpreting the guidelines from a legal perspective, it would be natural to distinguish 'shall' statements from 'should' statements and assume that the former are mandatory (prescriptive) and the latter discretionary.[13] Unfortunately, this method does not always yield coherent results. In accordance with article 12.2 of the FCCC, Annex I parties 'shall' communicate information on the policies and measures they have adopted to implement their commitments under article 4.[14] Under the guidelines, by contrast, parties 'should' give priority to reporting

---

11 FCCC (1999), *Guidelines on Reporting and Review*, section II, para. 8.
12 Ibid., section II, paras 27 and 40.
13 The FCCC parties are well aware of the different legal connotations of these two terms; see e.g. Clean Development Mechanism (2013), *CDM Validation and Verification Standard (v. 5.0)*, CDM-EB65-A04-STAN, para. 10 ('"Shall" is used to indicate requirements to be followed; "Should" is used to indicate that among several possibilities, one course of action is recommended as particularly suitable; "May" is used to indicate what is permitted').
14 See also FCCC (1999), *Guidelines on Reporting and Review*, section II, para. 13.

policies and measures that have the most significant impact on greenhouse gas emissions; moreover, they '[do] not have to report every policy and measure which affects GHG emissions'.[15] The 'should' in the rule creates room for debate about what the rule entails, despite the blanket 'shall' in the treaty itself.

A national communication 'should describe the overall policy context, including any national targets for greenhouse gas mitigation'. Strategies for sustainable development or other relevant policy objectives 'may also be covered'.[16] National mitigation targets are the most important element in the climate change regime, so the 'should' in this context surely must be interpreted as a mandatory 'shall'. The 'may also', however, seems genuinely discretionary. Sometimes the terminology changes in the course of the same paragraph: 'The presentation of each policy and measure *shall* include information on each of the subject headings listed below. ... Objectives of the policy or measure [this being a subject heading]: The description of the objectives *should* focus on the key purposes and benefits of the policies and measures'.[17] It seems to be a case of rule-making by negotiators rather than lawyers. In order to understand the content but also the legal status of these rules, context as well as the text of the decision must be taken into account.

Financial resources supplied by Annex I parties to non-Annex I parties, as required by article 4.3 of the FCCC, must be accounted for in national communications. It is a compulsory ('shall') requirement that only 'new and additional' financial support is listed and that parties clarify how they have determined that it is new and additional.[18] This is an area of considerable complexity (see Chapter 5), not least because of the large number of channels through which financial resources can be made available to developing countries.

Parties are required to report on technology transfer (article 4.5 of the FCCC) and to distinguish between transfer activities undertaken by the public sector and those undertaken by the private sector.[19] Other compulsory elements of reporting in national communications concern the support given to the 'enhancement of endogenous capacities' in non-Annex I parties,[20] information on actions relating to research and systematic observation,[21] and information on actions relating to education, training, and public awareness.[22]

The FCCC has a different set of guidelines for the national communications that are to be submitted by non-Annex I parties. The approach is markedly different, with the text of the relevant decision imbued with discretion and qualifications.[23] In applying these guidelines, non-Annex I parties 'should take into account their

---

15 Ibid., section II, para. 14.
16 Ibid., section II, para. 20.
17 Ibid., section II, para. 22, emphasis added.
18 Ibid., section II, para. 51.
19 Ibid., section II, para. 54.
20 Ibid., section II, para. 56.
21 Ibid., section II, para. 57.
22 Ibid., section II, para. 65.
23 See Lavanya Rajamani (2012), 'Developing Countries and Compliance in the Climate Regime', in *Promoting Compliance in an Evolving Climate Regime*, ed. Jutta Brunnée, Meinhard Doelle, and Lavanya Rajamani, p. 380.

development priorities, objectives and national circumstances'.[24] They are free to attempt to meet the higher standard of Annex I party reporting if they wish to.[25] The essence of the rule, however, is that they must report at a standard that is reasonably feasible for them. It is in principle an objective test, which is why the guidelines repeatedly invite non-Annex I parties to be specific about their financial, technical, and capacity needs.[26]

The main objective of non-Annex I national communications is to generate information relevant to the climate change regime that is transparent, consistent, and comparable. An additional aim is to provide input to the operating entity of the financial mechanism of the FCCC (originally the GEF) for the provision of financial support to non-Annex I parties to meet the 'agreed full costs' of complying with their obligations under the FCCC's article 12 (on communication of information related to the Convention's implementation).[27] The assumption here is that non-Annex I parties could improve on what is reasonably feasible for them to report on concerning their emissions and actions if they are properly aided to do so.[28]

For the foregoing reasons, non-Annex I national communications can take on characteristics of applications for financial assistance. Thus a non-Annex I national communication might go into detail about a reporting-related project for which the non-Annex I party is seeking financial support. At the same time, the reporting country is expected to disclose all information on resources it has received from the GEF, Annex I parties, or other bilateral or multilateral sources for the preparation of its national communication or other activities relating to climate change.[29] The guidelines repeat the text in article 4.7 of the FCCC, i.e. that the extent to which non-Annex I parties implement their commitment to communicate information to the other FCCC parties will depend on the implementation by Annex I parties of their own commitment to provide financial resources to those non-Annex I parties that need them.[30] This does not mean that the non-Annex I parties do not have an obligation to report. Rather, it means that because of their relatively low capacity, the information they communicate is unlikely to be fully transparent, consistent, or comparable without significant financial assistance (especially for least-developed countries) from Annex I parties. To put it another way, reporting is obligatory for non-Annex I parties where finance or other relevant support has been provided to facilitate it.

A non-Annex I national communication 'may include' information on features of the country's geography, climate, and economy which affect the government's

---

24 FCCC (2003), *Decision 17/CP.8, Guidelines for the Preparation of National Communications from Parties Not Included in Annex I to the Convention*, FCCC/CP/2002/7/Add.2, para. 1.
25 Ibid., para. 2.
26 Ibid., Annex, para. 49.
27 Ibid., Annex, para. 1.
28 The FCCC Secretariat facilitates assistance to non-Annex I parties in the preparation of their national communications: ibid., para. 3.
29 Ibid., Annex, paras 50–52.
30 Ibid., Annex, para. 27.

ability to pursue mitigation and adaptation measures.[31] The communication may provide information on the country's vulnerability to the adverse effects of climate change.[32] Information on mitigation and adaptation measures adopted or implemented may also be included.[33] It should be noted that adaptation is the main focus of the non-Annex I reporting guidelines, not mitigation. Non-Annex I parties are 'encouraged' to present evaluations of their strategies and measures for adaptation to climate change.[34] They are also encouraged to include information on capacity-building activities in their national communications.[35]

A non-Annex I party 'shall' inventorize its greenhouse gas emissions in its national communication. It is not obliged to provide its emission inventory separately (as Annex I parties are required to do). The first national communication for non-Annex I parties was to provide an inventory for 1994; the second was to inventorize emissions in the year 2000. Least-developed countries were permitted to report an inventory for any year, at their discretion.[36] Parties were to employ the IPCC's Revised 1996 Guidelines for National Greenhouse Gas Inventories for the production of their inventories. It is the same manual that Annex I parties were at the time required to use; however, non-Annex I parties were given greater flexibility in its application.[37]

I will now proceed to the rules concerning national communications submitted under the Kyoto Protocol. Each Protocol party in the Annex I group is obliged by article 5.1 of the Protocol to set up a 'national system' for the preparation of its greenhouse gas inventories.[38] (A national system was not a requirement under the FCCC.) A national communication prepared pursuant to the Protocol's guidelines must provide an account of the functioning of the state's national system. The information is to cover the institutional, legal, and procedural arrangements made to prepare the inventory, including the process for collecting activity data and selecting emission factors (also required under the FCCC), and must explain how the state developed and implemented a quality-assurance plan for the overall process (which goes beyond the FCCC requirements).[39]

The Kyoto Protocol created a variety of emission allowances specific to its three market mechanisms.[40] Trading in emission allowances under these so-called

---

31 Ibid., Annex, para. 3.
32 Ibid., Annex, paras 28–29.
33 Ibid., Annex, para. 26.
34 Ibid., Annex, para. 35.
35 Ibid., Annex, para. 46.
36 Ibid., Annex, para. 7.
37 Ibid., Annex, paras 8–9.
38 Kyoto Protocol (2005), *Decision 19/CMP.1, Guidelines for National Systems under Article 5, Paragraph 1, of the Kyoto Protocol*, FCCC/KP/CMP/2005/8/Add.3.
39 Kyoto Protocol (2005), *Decision 15/CMP.1, Guidelines for the Preparation of the Information Required under Article 7 of the Kyoto Protocol*, FCCC/KP/CMP/2005/8/Add.2, Annex, para. 30.
40 For a general introduction, see Erik Haites and Farhana Yamin (2004), 'Overview of the Kyoto Mechanisms', 5 (1) *International Review for Environmental Strategies* 199.

flexibility mechanisms has required the establishment of national registries linked to a centralized international registry.[41] In its national communication an Annex I party is to account for the performance of its registry and the extent of its compliance with the technical standards for data exchange between registry systems. The state's registry must be enabled for data exchange with other national registries, the Clean Development Mechanism registry (maintained by the CDM Executive Board), and the International Transaction Log (maintained by the FCCC Secretariat).[42]

In accordance with the Protocol, trading of emission allowances 'shall be supplemental to domestic actions', and while this in itself is an ill-defined threshold, an Annex I party must be able to show that its mitigation strategy does not over-rely on the flexibility mechanisms but is instead mostly accounted for by domestic mitigation action.[43] Accordingly, Annex I parties are to report on domestic legislative arrangements and administrative and enforcement procedures, implemented with the aim of achieving each party's 'quantified emission limitation and reduction commitment' under article 3 of the Protocol.[44]

The obligation upon Annex I parties to transfer finance and technology to non-Annex I parties for mitigation and adaptation was not further developed in the Kyoto Protocol. Nonetheless, the Protocol's reporting guidelines oblige each Annex I party to account in its national communication for any new and additional resources made available to developing countries to assist them with reporting, mitigation, and adaptation.[45]

I will pause here to emphasize that the obligations discussed in this chapter have to do with the reporting of actions rather than with the actions themselves. A reporting requirement is met if it addresses the action area, even if the state has been inactive in that area. States will always have emissions to report, of course, so there could never be inaction in this area. If a state has introduced no new mitigation measures or made no financial transfers in support of developing-country action, it must report that fact, and by doing so will have complied with its reporting obligations, if not with any applicable substantive obligation. The only exception to the reporting/action distinction is that, as discussed above, the Protocol requires Annex I parties to set up specific administrative structures, namely national systems and emission-allowance registries. Reporting on certain matters will not be accepted unless it has been produced within these structures. In this case only, compliance with reporting obligations presupposes substantive action, but the substance concerns the implementation of methods to improve the quality of reporting.

---

41 Kyoto Protocol (2005), *Decision 19/CMP.1*; and Kyoto Protocol (2005), *Decision 12/CMP.1, Guidance Relating to Registry Systems under Article 7, Paragraph 4, of the Kyoto Protocol*, FCCC/KP/CMP/2005/8/Add.2.
42 Kyoto Protocol (2005), *Decision 15/CMP.1*, Annex, para. 32.
43 Kyoto Protocol (2005), *Decision 2/CMP.1, Principles, Nature and Scope of the Mechanisms Pursuant to Articles 6, 12 and 17 of the Kyoto Protocol*, FCCC/KP/CMP/2005/8/Add.1, para. 1.
44 Kyoto Protocol (2005), *Decision 15/CMP.1*, Annex, para. 37.
45 Ibid., Annex, para. 41.

National reporting and assessment 37

While it might be said that reporting is only a procedural requirement and thus should be discussed after the rules on substantive action, from a legal point of view the international regime's reporting rules are logically prior to those on substance, as they must be complied with first, before it can be known whether the obligations on substantive matters have been complied with. A state's ability and motivation to comply with substantive rules would be severely diminished where reporting rules are not complied with by the state itself or by other states in the first place. The substantive rules of international climate change law are reviewed in Chapters 4 and 5.

2.2.3 Greenhouse gas inventories

Article 4.1 of the FCCC requires all states to 'Develop, periodically update, [and] publish ... national inventories of anthropogenic emissions by sources and removals by sinks of all greenhouse gases ... using comparable methodologies'. Several decisions[46] and guidance documents[47] have elaborated the obligations of the parties to report their emissions. Annex I inventories must be reported annually[48] whereas non-Annex I emission information is, as noted earlier, incorporated into national communications and updated only when a new national communication is prepared.[49] Non-Annex I emissions are thus reported relatively infrequently.[50]

---

46 FCCC Subsidiary Body for Scientific and Technical Advice (2004), *Guidelines for the Preparation of National Communications by Parties Included in Annex I to the Convention, Part I: UNFCCC Reporting Guidelines on Annual Inventories (Following Incorporation of the Provisions of Decision 13/CP.9)*, FCCC/SBSTA/2004/8; FCCC Subsidiary Body for Scientific and Technical Advice (2006), *Updated UNFCCC Reporting Guidelines on Annual Inventories Following Incorporation of the Provisions of Decision 14/CP.11*, FCCC/SBSTA/2006/9; and FCCC (2013), *Decision 24/CP.19, Revision of the UNFCCC Reporting Guidelines on Annual Inventories for Parties Included in Annex I to the Convention*, FCCC/CP/2013/10/Add.3.
47 IPCC (1996), *Revised 1996 IPCC Guidelines for National Greenhouse Gas Inventories*; IPCC (2000), *Good Practice Guidance and Uncertainty Management in National Greenhouse Gas Inventories*; IPCC (2003), *Good Practice Guidance for Land Use, Land-Use Change and Forestry*; and IPCC (2006), *IPCC Guidelines for National Greenhouse Gas Inventories* (combining LULUCF and agriculture into a single Agriculture, Forestry, and Other Land-Uses (AFOLU) sector). Most of these manuals are technical in nature, extensive in scope, and complex in content, reflecting the complexity of the relevant issues and processes.
48 FCCC (1998), *Decision 11/CP.4, National Communications from Parties Included in Annex I to the Convention*, FCCC/CP/1998/16/Add.1, para. 2.b; and FCCC (2002), *Decision 18/CP.8, Guidelines for the Preparation of National Communications by Parties Included in Annex I to the Convention, Part I: UNFCCC Reporting Guidelines on Annual Inventories*, FCCC/CP/2002/7/Add.2, para. 2.
49 In 2011 the FCCC decided that the IPCC (1996), *Revised Guidelines* in conjunction with the IPCC (2000), *Good Practice Guidance* and IPCC (2003), *Good Practice Guidance for LULUCF* should be used by developing countries for estimating and reporting anthropogenic emissions and removals (FCCC (2009), *Decision 4/CP.15, Methodological Guidance for Activities Relating to Reducing Emissions from Deforestation and Forest Degradation*, FCCC/CP/2009/11/Add.1).
50 See Table 4.4, in Chapter 4.

According to the FCCC's reporting guidelines on annual inventories,[51] an emission inventory 'should' be transparent (transparency here means that the assumptions and methodologies used to produce the inventory should be explained), consistent (including consistent with the state's inventories from earlier years), comparable among Annex I parties, complete (i.e. all sources and sinks and all gases should be covered), and accurate, with the uncertainties reduced as far as practicable.[52] We may deduce from the context that the 'should' prefacing these important qualities is meant as a mandatory 'shall'. The applicable FCCC decisions switch between 'shall' and 'should', at times apparently randomly.

In striving to attain the above standards, Annex I parties 'shall' use the IPCC's various guidance documents.[53] However, they may also use 'national methodologies' which they consider as better reflecting their national circumstances, provided that these alternatives are 'compatible' with IPCC guidance and are documented and scientifically based.[54] For most source categories of emissions, the IPCC provides only a default methodology, including default emission factors and in some cases even default activity data. It is, therefore, preferable for Annex I parties to use their own national emission factors and activity data, where available, for greater accuracy.[55] Up to a point, the accuracy to be attained in this matter is discretionary.

Similarly, the completeness rule requires less than absolute completeness. A party may consider that a disproportionate amount of effort would be required to collect emission data for a greenhouse gas from a given activity category in a situation where the amount emitted is likely to be insignificant compared with the country's aggregate emissions. The rules define an insignificant amount as a likely level of emissions below 0.05 per cent of the national total, but not exceeding 500 kt CO_2 eq.[56]

Emissions are to be reported at the most disaggregated level for each category, although a higher level of aggregation may be required (and is allowed) to protect confidential commercial and military information.[57] In other respects, where data gaps exist in inventories, a rationale for them should (must) be presented in a transparent manner. Parties are required to indicate the emission sources not considered in their inventories yet included in the IPCC's guidance documents, and explain the reason for the exclusion.[58]

A national greenhouse gas inventory consists of tables of numerical data and a separate, qualitative, national inventory report (NIR). Reporting rules apply to the NIR as well as to the data tables. The NIR's purpose is to contribute to the transparency of the data. For this, it should contain sufficiently detailed

---

51 FCCC (2002), *Decision 18/CP.8*, para. 1. The annex to this decision containing the guidelines is in fact a separate document; see FCCC (2002), *Guidelines on Reporting and Review*, FCCC/CP/2002/8.
52 FCCC (2002), *Guidelines on Reporting and Review*, Reporting Guidelines, para. 2.
53 Ibid., Reporting Guidelines, para. 9.
54 Ibid., Reporting Guidelines, para. 10.
55 Ibid., Reporting Guidelines, para. 12.
56 FCCC (2013), *Decision 24/CP.19*, Annex, para. 37(b).
57 FCCC (2002), *Guidelines on Reporting and Review*, Reporting Guidelines, para. 27.
58 Ibid., Reporting Guidelines, para. 28.

information to enable the inventory to be reviewed.[59] In the NIR, parties are to report, among other things, on the uncertainties affecting their emission estimates. This information is meant to help parties prioritize their efforts to improve the accuracy of their inventory in future submissions.[60] The NIR must also include a description of the institutional arrangements followed for inventory preparation[61] and the inventory quality-assurance plan implemented during the preparation process.[62]

As with national communications, the Kyoto Protocol's inventory requirements differ somewhat from the FCCC's. Annex I parties to the Protocol must comply with an additional set of rules on the preparation of emission reports. One reason for the additional reporting is that article 3.3 of the Protocol requires emissions and removals from land-use activities (afforestation, reforestation, and deforestation) to be counted toward a party's 'assigned amount' of emissions for a commitment period. There is, of course, no parallel to this in the FCCC. Emissions and removals from another set of land-use activities, under Protocol article 3.4 (forest management, cropland management, grazing-land management, and revegetation), may be counted toward the assigned amount if a country so chooses.[63]

The Protocol's flexibility mechanisms entail that a state's inventory is no longer constituted exclusively of the emissions and removals taking place in the party's own territory. The inventory may include emission allowances formerly held by other Annex I parties (AAUs and ERUs) or generated in non-Annex I parties through the CDM (CERs). This information must be reported in the national inventory report.[64] Trading information held in registries must be reconciled to ensure that a transferred allowance is cancelled in the account of origin. Transaction discrepancies that are not cleared up by the party concerned could raise a 'question of implementation' (see next section as well as Chapter 3).[65]

The Protocol is operationalized through commitment periods, which create their own special reporting obligations since an assigned amount covers a whole commitment period not any particular reporting year. An NIR for the final year of a commitment period must therefore address the critical questions of whether the party has complied with its assigned amount and whether it has done so substantially through a domestic mitigation effort.[66]

---

59 Ibid., Reporting Guidelines, para. 38.
60 Ibid., Reporting Guidelines, para. 32.
61 Ibid., Reporting Guidelines, para. 41.
62 Ibid., Reporting Guidelines, para. 17.
63 Kyoto Protocol (2005), *Decision 15/CMP.1*, Annex, para. 5. From the second commitment period onward, art. 3.4 activities *must* be counted toward the assigned amount.
64 Ibid., Annex, para. 10.
65 Ibid., Annex, para. 17.
66 Ibid., Annex, para. 20.

2.2.4 Expert review

As previously discussed, the FCCC requires all parties to have in place policies and measures to mitigate climate change and facilitate adaptation. Annex I parties are to report in detail on the implementation and impact of their policies and measures, and non-Annex I parties must do the same to the extent of their capacity.

The FCCC has established a review system for state reports. It is limited to the reports of Annex I parties. Their national communications are subject to 'in-depth review' whereas their inventories must undergo 'technical review'. The Kyoto Protocol has added to the review procedures, with the so-called 'periodic' reviews of national communications and 'annual reviews' of inventories. All these reviews are carried out by Expert Review Teams.

The FCCC's 'in-depth review' procedure dates back to 1995 and has not substantially changed since. Each national communication is to be reviewed, within a year of submission, by an Expert Review Team.[67] ERTs were at the time constituted of persons selected by the FCCC Secretariat from lists of experts nominated by parties and intergovernmental organizations. (IGOs no longer nominate ERT members.) The Secretariat is to ensure 'a balance of skills and expertise, of environmental and developmental perspectives and the necessary geographical balance among team members'.[68] A national of the state under review is prevented from joining an ERT charged with that state's review.

The in-depth review process is to provide 'a thorough and comprehensive technical assessment of the implementation of the Convention commitments' by Annex I parties, individually but also in aggregate.[69] The review reports are to provide the Conference of the Parties with 'accurate, consistent and relevant information ... to assist it in carrying out its responsibilities, [namely] To assess the implementation of the Convention by the Parties ... and the extent to which progress towards the objective of the Convention is being achieved'.[70] In-depth review reports are to be written in 'non-confrontational language'.[71] It thus seems that the FCCC review process was conceived of at least initially as being about improving information quality rather than about 'compliance' in a legal sense. Indeed, the FCCC does not incorporate a formal compliance system.

The tasks of an Expert Review Team when conducting an in-depth review are given in rather broad terms. An ERT is to: 'review' qualitative information and quantitative data in national communications; 'review' state policies and measures; 'assess' the submitted information against FCCC commitments, including those on the provision of financial support to non-Annex I parties; and 'assess' the extent of the state's progress toward limiting its greenhouse gas emissions in accordance with the Convention's article 2 objective.[72] In-depth review reports by ERTs are

---

67 FCCC (1995), *Decision 2/CP.1, Review of First Communications from the Parties Included in Annex I to the Convention*, FCCC/CP/1995/7/Add.1, para. 2.a.
68 Ibid., para. 4.b.
69 Ibid., Annex I.
70 Ibid., Annex I.
71 Ibid., para. 2.d.
72 Ibid., Annex II.

National reporting and assessment 41

to be forwarded to the Subsidiary Body on Implementation for 'consideration'.[73] The FCCC's review procedure for national communications concludes at this point without further elaboration.

An annual 'thorough and comprehensive technical assessment of national inventories'[74] of Annex I parties became a requirement under the FCCC in 2003. A specific procedure applies to this type of review.[75] Inventories are to be reviewed by Expert Review Teams against the standard of being 'transparent, documented, consistent over time, complete, comparable, assessed for uncertainties, [and] subject to quality control and quality assurance'.[76] Institutional arrangements for inventory development and management must also be reviewed by the ERT.[77]

The 'technical review' may be centralized, in which case the Expert Review Team convenes at the Secretariat's offices in Bonn to review the inventory information. Alternatively, it may be an in-country review, involving a visit by an ERT to the country under review. An Annex I party must undergo an in-country technical review once every five years.[78]

An ERT must check that the national inventory is consistent with the applicable guidelines. It must 'identify any departure from these requirements'.[79] This would be an adequate instruction if the reporting guidelines clearly stated what is mandatory and what is not in a greenhouse gas inventory. But that is far from being the case, as I indicated earlier in this chapter. A revision of the inventory review guidelines in 2013 did not bring any clarity in this respect. Some examples of the language in the guidelines will help to underscore the problem: 'each Annex I Party should prepare national annual GHG inventories in a timely manner in accordance with these reporting guidelines ... Annex I Parties shall use the methodologies provided ... Annex I Parties shall estimate and report the individual and cumulative percentage contributions from key categories to their national total ... Each Annex I Party should implement and maintain national inventory arrangements ... Annex I Parties shall quantitatively estimate the uncertainty which should also be qualitatively discussed in a transparent manner in the NIR ... the underlying AD [activity data] and EFs [emission factors] should be obtained and used in a consistent manner ... Recalculations should ensure the consistency of the time series and shall be carried out to improve accuracy ... Annex I Parties should document and report the methodologies used for the entire time series ... Where methodological or data gaps in inventories exist, information

---

73 Ibid., para. 2.e.
74 FCCC (2002), *Guidelines on Reporting and Review*, Technical Review, para. 1.
75 FCCC (2002), *Decision 19/CP.8, Guidelines for the Technical Review of Greenhouse Gas Inventories from Parties Included in Annex I to the Convention*, FCCC/CP/2002/7/Add.2, paras 1–2. The annex to this decision containing the guidelines is in fact a separate document: FCCC (2002), *Guidelines on Reporting and Review*.
76 Preface to IPCC (2000), *Good Practice Guidance*.
77 FCCC (2002), *Guidelines on Reporting and Review*, Technical Review, para. 22.
78 Ibid., Technical Review, paras 18–19. The normal size of an ERT is 12 experts for centralized reviews and six experts for in-country reviews: ibid., Technical Review, para. 32.
79 Ibid., Technical Review, para. 21.

on these gaps should be presented in a transparent manner'.[80] In the absence of a formal compliance system that can raise allegations of state non-compliance with treaty obligations, the loose language of the guidelines possibly was not seen as a problem by the FCCC parties. In such a context of 'no consequences' it is not critical that an ERT is able to draw a hard line between mandatory and discretionary language. The loose language is in itself evidence that an ERT's review of the Annex I greenhouse gas inventory was to be facilitative, not confrontational, in spirit.

Additional grounds for this argument can be found in the review procedures. The ERT is to examine the reported inventory information in an 'open and facilitative manner'.[81] There is to be no 'political' judgement.[82] The ERT is to identify areas for further improvement and advise the state on how such improvements could be made.[83] At all stages of the inventory review process, an Annex I party is to have the opportunity to clarify issues or provide additional information. Every effort is to be made by the ERT to reach agreement with the party on the content of a review report prior to its publication. Where the ERT and the party are unable to agree on an issue, the party may provide an explanation, which is to be attached to the ERT's report.[84] These provisions together stand for a facilitative, non-confrontational, process.

On the basis of what has been said so far, one is in a position to predict that, in the Kyoto Protocol context, where a formal compliance system does exist and where the same FCCC guidelines for inventory preparation apply for the most part, problems of interpretation for ERTs of the ambiguous guideline instructions cannot be avoided.

I will now briefly discuss the review procedures that apply to national communications and greenhouse gas inventories under the Kyoto Protocol ('periodic reviews' and 'annual reviews', respectively).[85] Both types of review are conducted at the same time as the corresponding FCCC review.[86] The basic review tasks for Expert Review Teams in connection with periodic and annual reviews are variations on the tasks I outlined above, and need not be further detailed here.[87] I will focus on the one great difference between the two systems: Protocol reviews share the aim of FCCC reviews (namely, a thorough and comprehensive technical assessment of all aspects of implementation), but have a unique additional aim:

---

80 FCCC (2013), *Decision 24/CP.19*, Annex.
81 FCCC (2002), *Guidelines on Reporting and Review*, Technical Review, para. 2.
82 Ibid., Technical Review, para. 39.
83 Ibid., Technical Review, para. 21.
84 Ibid., Technical Review, para. 5.
85 Kyoto Protocol, art. 8.
86 Kyoto Protocol (2005), *Decision 22/CMP.1, Guidelines for Review under Article 8 of the Kyoto Protocol*, FCCC/KP/CMP/2005/8/Add.3, Annex, para. 1.
87 For periodic reviews of national communications under the Kyoto Protocol, see ibid., Annex, paras 131–140; for annual reviews of greenhouse gas inventories, see ibid., Annex, paras 50–117.

'The expert review team shall ... identify any potential problems in, and factors influencing, the fulfilment of commitments.'[88]

A 'potential problem' arises where an ERT has reason to believe that the Protocol party under review has not fully adhered to the reporting guidelines on national communications or inventories. The ERT must seek to get to the bottom of any such problem. It must question the party on the apparent problem and offer advice on how it should be corrected if it is real.[89] A state party is expected to respond to all ERT questions and requests for clarifying information relating to the 'potential problem'—and it is also expected to quickly correct it.[90] If the problem is not resolved, one of two things can happen. The ERT may designate the problem as a 'question of implementation', but only if the problem 'pertain[s] to language of a mandatory nature in these guidelines influencing the fulfilment of commitments'. Otherwise, it must do no more than note the existence of the problem in its final review report.[91] 'A question of implementation' is not dealt with under the review procedure itself, but is directed to the Protocol's Compliance Committee to be dealt with under the compliance procedure (Chapter 3).

The category of a 'question of implementation' under the Kyoto Protocol's review system brings to a head the issue I have been drawing attention to about the normatively imprecise language of the review guidelines. I will return to the issue in the next chapter.

I close this section with a comment on the annual inventory review ('annual review') under the Protocol. It requires an Expert Review Team to review two elements that go to a state party's capacity to report accurately on its emissions and emission-allowance holdings: the national system and the registry. The ERT must assess the 'capacity' and 'overall organization' of the national system, and the 'adequacy of its institutional, legal and procedural arrangements to produce an inventory'.[92] The national registry must be reviewed with a similar thoroughness to ensure that it functions in such a way that the information it holds relating to the issuance, cancellation, retirement, transfer, acquisition, and replacement of emission allowances is accurate.[93] Shortcomings in national systems or registries could give rise to questions of implementation,[94] and indeed the first ever question of implementation to be raised by an ERT was about a national system.[95]

2.2.5 New directions: Biennial reports and IAR/ICA

There is a clear trend toward enhanced reporting by all FCCC parties. An example of this is a 2010 decision of the FCCC on national communications from

---

88 Ibid., Annex, para. 4.
89 Ibid., Annex, para. 7.
90 Ibid., Annex, para. 6.
91 Ibid., Annex, para. 8.
92 Ibid., Annex, paras 96 and 109.
93 Ibid., Annex, paras 88 and 110.
94 See ibid., Annex, paras 106 and 117.
95 On the case of Greece before the Compliance Committee, see Chapter 3.

non-Annex I parties for the 2013–2020 period: they are henceforth to be submitted every four years.[96] In addition, there is to be biennial reporting on the mitigation pledges made by FCCC parties (both developed and developing) for the 2013–2020 period,[97] and these reports are to undergo a degree of scrutiny through the new processes of International Assessment and Review and International Consultation and Analysis.[98] It should be recalled that prior to this point in time the FCCC had not developed provisions for reviewing progress toward the achievement of emission-reduction targets, and indeed had not set any such targets for any year in the period 2001–2012. (For the period up to 2000, when what I have called the specific mitigation rule applied, see Chapter 4.)

The first biennial reports from Annex I parties on their 2013–2020 mitigation pledges were due in January 2014.[99] Those from developing countries were due in December of the same year. Least-developed countries and small-island developing states were not required to report on their so-called 'nationally appropriate mitigation actions' (NAMAs), but were invited to do so at their discretion.[100] (The terminology applicable to the 2013–2020 period is discussed in more detail in Section 4.1.3 in Chapter 4.)

The IAR and ICA procedures have been developed for the review of the biennial reports of Annex I and non-Annex I parties, respectively, in the period to 2020. IAR commenced two months after the submission of the first round of Annex I biennial reports.[101] ICA for the non-Annex I reports was to commence in mid-2015. The costs of ICA are to be met from additional financial contributions by Annex I parties.[102]

The objective of IAR is to review the progress made by an Annex I party toward achieving its pledged emission reductions (its 'quantified economy-wide emission-reduction target') for the period through to 2020 and to assess its provision of financial support for mitigation to developing countries.[103]

---

96 FCCC (2010), *Decision 1/CP.16, The Cancun Agreements: Outcome of the Work of the Ad Hoc Working Group on Long-Term Cooperative Action under the Convention*, FCCC/CP/2010/7/Add.1, para. 60.b. Compliance with this requirement is subject to 'the prompt provision of financial resources to cover the agreed full costs' of preparing national communications.
97 See ibid., para. 36.
98 Ibid., paras 40.a and 60.c. These are 'should' rather than 'shall' provisions.
99 FCCC (2011), *Decision 2/CP.17, Outcome of the Work of the Ad Hoc Working Group on Long-Term Cooperative Action under the Convention*, FCCC/CP/2011/9/Add.1, paras 12–13.
100 Ibid., para. 41.a; FCCC (2013), *Decision 21/CP.19, General Guidelines for Domestic Measurement, Reporting and Verification of Domestically Supported Nationally Appropriate Mitigation Actions by Developing Country Parties*, FCCC/CP/2013/10/Add.2, para. 2.
101 FCCC (2011), *Decision 2/CP.17*, paras 23 and 25.
102 Ibid., paras 58–59; and FCCC (2013), *Decision 20/CP.19, Composition, Modalities and Procedures of the Team of Technical Experts under International Consultation and Analysis*, FCCC/CP/2013/10/Add.2, para. 6.
103 FCCC (2013), *Decision 23/CP.19, Work Programme on the Revision of the Guidelines for the Review of Biennial Reports and National Communications, Including National*

Another novel element of the procedure is that it is intended to shed light on the comparability of the mitigation efforts of Annex I parties.[104] (As explained in Chapter 4, the pledges for the 2013–2020 period are bottom-up and essentially ad hoc. They are not referenced to a global emission budget, and by 2014, when IAR went into effect, it was still not known how they compared for effort.) Whether IAR has the capacity to answer this question is another matter.

IAR is conducted under the oversight of the SBI, in two steps: a technical review of biennial reports by Expert Review Teams, and a 'multilateral assessment' of the review reports and related information within the framework of SBI meetings. The review by ERTs is to identify any 'potential issues' (different from the Kyoto Protocol's 'potential problems') arising from a biennial report.[105] These reports are subject to methodological and reporting requirements. Departures from the requirements could give rise to a finding about 'potential issues'.[106] Once transparency and accuracy of reporting are established through technical review by an ERT, the substance of the state's actions and its relationship to the actions of other states can be considered in the SBI forum.

Multilateral assessment of the biennial reports of Annex I parties is to be held during SBI sessions.[107] The assessment must consider the reviewed state's reported emissions and compare them with its quantified economy-wide emission-reduction target, the assumptions and methodologies related to the attainment of that target, and the state's progress toward the achievement of the target.[108] Information on the provision of financial support to developing countries is *not* assessed multilaterally and is only subject to technical review by ERTs.[109] Prior to the SBI multilateral assessment meeting, any party may submit written questions to the Annex I party concerned, and that party must endeavour to respond to them. The record of such an exchange is be published on the FCCC website (potentially naming and shaming underperforming states). During the actual SBI session, all parties may participate in the assessment without prior written notice. The party under review may make a brief oral presentation, which is to be followed by oral questioning by the parties in attendance.[110] The SBI is to forward its conclusions based on the record of the meeting to 'relevant bodies under the Convention as appropriate'.[111]

Inventory Reviews, for Developed Country Parties, FCCC/CP/2013/10/Add.2, Annex, para. 58.
104 Ibid., Annex, para. 6.
105 Ibid., Annex, para. 10.
106 Ibid., Annex, para. 59.b. For the technical review procedure and guidelines, see FCCC (2011), *Decision 2/CP.17*, Annex I, as well as FCCC (2013), *Decision 23/CP.19*, Annex, paras 61–71.
107 FCCC (2011), *Decision 2/CP.17*, Annex II, para. 9.
108 Ibid., Annex II, para. 5.
109 On this, see Yulia Yamineva and Kati Kulovesi (2013), 'The New Framework for Climate Finance under the United Nations Framework Convention on Climate Change: A Breakthrough or an Empty Promise?', in *Climate Change and the Law*, ed. Erkki J. Hollo, Kati Kulovesi, and Michael Mehling, pp. 215–216.
110 FCCC (2011), *Decision 2/CP.17*, Annex II, para. 10.
111 Ibid., Annex II, para. 12.

The ICA procedure for non-Annex I parties, also under the SBI, is differently constituted: a 'technical analysis' (as opposed to a technical review) of non-Annex I biennial reports, followed by a 'facilitative sharing of views'.[112] The technical analysis is to be carried out by a team of technical experts (or TTE, to be distinguished from an ERT), composed of persons nominated to the FCCC Roster of Experts who have completed a special training programme under the Consultative Group of Experts.[113] The CGE,[114] which is itself composed of experts from the Roster of Experts, is to provide ongoing advice to non-Annex I parties to facilitate the work of their 'national technical teams' in the preparation of biennial reports.[115] The majority of TTE experts are to be from developing countries.[116] The technical analysis is to aim to 'increase transparency of mitigation actions and their effects' and identify capacity-building needs in order to improve reporting.[117] The technical analysis is to result in a 'summary' report (to be distinguished, again, from the standard, in-depth, ERT report).[118] The summary report is to be finalized in consultation with the party concerned and submitted to the SBI.[119]

ICA provides for a highly attenuated multilateral component. After final summary reports are received by the SBI, the body is to convene workshops at regular intervals, open to all FCCC parties, 'for the facilitative exchange of views' on the non-Annex I biennial reports and final summary reports. Up to five non-Annex I parties are to be discussed at each workshop. The procedure specifies that a workshop is to run for 1–3 hours. Parties may submit written questions in advance.[120] At the workshop, a brief presentation by the non-Annex I party (or parties) concerned on its biennial report is to be followed by an oral question-and-answer session.[121] The outcome of the ICA is to be another summary report and a record of the 'facilitative sharing of views'.[122] These are to be made public on the FCCC website.[123] ICA is to be 'non-intrusive, non-punitive and respectful of national sovereignty [of non-Annex I parties], and … does not include … discussion about the appropriateness of [their] domestic policies and measures'.[124]

---

112 Ibid., Annex IV, para. 3.
113 FCCC (2013), *Decision 20/CP.19*, Annex, para. 3.
114 The CGE's full name is Consultative Group of Experts on National Communications From Parties Not Included in Annex I to the Convention.
115 FCCC (2013), *Decision 19/CP.19, Work of the Consultative Group of Experts on National Communications from Parties Not Included in Annex I to the Convention*, FCCC/CP/2013/10/Add.2, Annex, paras 2.c and 4.
116 FCCC (2013), *Decision 20/CP.19*, Annex, para. 5. In ERTs, by contrast, developed and developing countries are equally represented.
117 Ibid., Annex, para. 15.
118 Ibid., Annex, para. 8. For the technical analysis procedure and guidelines, see FCCC (2011), *Decision 2/CP.17*, Annex III.
119 FCCC (2011), *Decision 2/CP.17*, Annex IV, para. 4.
120 Ibid., Annex IV, para. 6.
121 Ibid., Annex IV, para. 7.
122 Ibid., Annex IV, para. 8.
123 Ibid., Annex IV, para. 5.
124 FCCC (2013), *Decision 20/CP.19*, Annex, para. 1.

Rajamani observes that ICA is essentially confined 'to the non-threatening realm of technical experts'.[125] According to Vihma, however, ICA represents an unprecedented 'internationalization' of the reporting requirements of developing countries:

> Since the launch of the UN climate regime the developing countries have faced virtually no transparency requirements ... the National Communications have not been regular, they have not been designed in accordance with international guidelines, and they have been allowed to use ancient data. Comparing this long-time status quo to the biennial reporting with 4-year-old data [in the national communications] and an ICA procedure ... shows a significant step forward in the hardness of the 2013–2020 climate regime.[126]

Hard or not, the 'enhanced' reporting scheme for the pledge period does not provide for a punitive response against a party that is failing to comply with the requirements for biennial reporting, or indeed with its mitigation targets or NAMAs. The multilateral components of IAR and ICA suggest an intention to expose states to peer pressure. In this, they stop well short of the compliance system under the Protocol. Oberthür has expressed concern about the design elements of the FCCC's new system:

> The multilateral phases of both IAR and ICA are clearly facilitative. They primarily aim at enhancing transparency. ... Neither multilateral phase foresees findings or concrete results directed at the party in question or at others (e.g. institutions providing financial assistance). ... The independence and institutional capacity of both the IAR's multilateral assessment and the ICA's facilitative sharing of views remain seriously lacking. The multilateral phases of both procedures are not conducted by a dedicated committee but in the context of the SBI, which forms part of a bigger political process.[127]

2.3 State compliance in practice

There are two main sources from which to gauge compliance with the FCCC's reporting rules. First, the FCCC Secretariat has prepared so-called compilation-and-synthesis reports on five rounds of national communications from Annex I parties and on one round from non-Annex I parties. Second, Expert Review Teams

---
125 Lavanya Rajamani (2012), 'Developing Countries and Compliance', p. 381.
126 Antto Vihma (2013), 'Analyzing Soft Law and Hard Law in Climate Change', in *Climate Change and the Law*, ed. Erkki J. Hollo, Kati Kulovesi, and Michael Mehling, p. 161.
127 Sebastian Oberthür (2014), 'Options for a Compliance Mechanism in a 2015 Climate Agreement', 4 (1–2) *Climate Law* 30, pp. 42–43. See also Geir Ulfstein (2012), 'Depoliticizing Compliance', in *Promoting Compliance in an Evolving Climate Regime*, ed. Jutta Brunnée, Meinhard Doelle, and Lavanya Rajamani, p. 428.

carry out in-depth reviews of the communications and inventories of Annex I parties (only). I will consider these two sources in turn.

2.3.1 Assessment of national communications

Annex I parties submitted national communications, as required, every four to five years. The fifth round was completed in 2010 and the sixth in early 2014.

The first compilation-and-synthesis report of national communications from Annex I parties is from 1994. It was based on the communications of 15 states.[128] Because the Secretariat is a neutral body with no assessment mandate of its own, only the mildest critique of state performance would be expected from a compilation-and-synthesis report. The 1994 report implied that state compliance with reporting rules had been satisfactory. The one problem highlighted has proven to be an extremely difficult one to solve. The report noted that the level of 'new and additional' financial resources provided to non-Annex I parties 'cannot be determined on the basis of the [national] communications as there is no agreed benchmark against which to measure this'; moreover, 'it was not possible to aggregate the reported resource flows owing to a lack of comparable data.'[129] The next report by the Secretariat, based on the communications of 31 Annex I parties (including the 15 of the first report), highlighted problems with the reporting of 'policies and measures' to reduce emissions. The Secretariat noted that information on the effects of specific measures 'was often sketchy' or based on hidden assumptions; and the cost-effectiveness of the measures was rarely discussed, making it difficult to evaluate what measures were the most (cost) effective.[130] The Secretariat's complaint about the non-transparency of state reporting on policies and measures recurs in all the reports that followed. In later waves of national communications, by which time states had begun to grapple with counterfactual analyses (especially with emission projections 'with' and 'without' additional measures), many methodological difficulties, and thus further problems with transparent reporting, became evident.[131]

The Secretariat on several occasions suggested that reporting shortcomings could be addressed through improved reporting guidelines, implying that the

---

128 It was the first instalment of a two-part report.
129 FCCC Interim Secretariat (1994), *First Compilation and Synthesis of First National Communications from Annex I Parties*, A/AC.237/81, paras 16–17; see also FCCC Subsidiary Body for Implementation (2011), *Compilation and Synthesis of Fifth National Communications from Annex I Parties: Executive Summary*, FCCC/SBI/2011/INF.1, para. 34. See Chapter 5 for more detail on this problem.
130 FCCC Secretariat (1996), *Second Compilation and Synthesis of First National Communications from Annex I Parties: Executive Summary*, FCCC/CP/1996/12, para. 29.
131 FCCC Subsidiary Body for Implementation (2003), *Compilation and Synthesis of Third National Communications from Annex I Parties: Executive Summary*, FCCC/SBI/2003/7, para. 27.

states themselves were not always to blame.¹³² There were, however, periods when a significant number of Annex I parties were not reporting in a timely manner, and for this, presumably, they were themselves to blame. For example, only nine parties submitted their second national communication by the due date, and five parties still had not submitted it eighteen months after the expiration of the deadline.¹³³ Eventually, though, all reports were submitted.

While the Secretariat's compilation-and-synthesis reports are not particularly incisive and generally avoid naming individual countries, the overall impression they impart is that Annex I parties have consistently complied with the rules on the reporting of their actions. The Secretariat makes a strong case for there having been an incremental improvement in national communications between the first round and the fifth (a compilation-and-synthesis report on the sixth round was not available at the time of writing). There is indeed no evidence of systematic avoidance of reporting obligations. For a deeper country-by-country analysis, one must turn to the in-depth ERT reviews of the national communications of Annex I parties. It is a point I return to below.

As there has been only one complete round of national communications from non-Annex I parties, the Secretariat has produced only one compilation-and-synthesis report in that context. Because no submission deadline was set for developing countries, their first national communications trickled in over the course of seven years, and the Secretariat's compilation-and-synthesis report is in six parts. It is an indication of the gulf between developed- and developing-country reporting practices.¹³⁴

The first instalment of the Secretariat's report was in 1999, by which time eleven non-Annex I national communications had been received. The Secretariat stated that 'a dominant theme ... is the need for better quality of data, ... financial resources and technical expertise', as well as institutional development.¹³⁵ Developing countries were openly admitting that their reports were not up to standard in all respects, and also arguing that substantial assistance from Annex I parties was needed to address the problem. Nevertheless, 'In their use of the guidelines, Parties covered in considerable detail the areas, sectors and activities on which information was requested.'¹³⁶ By the following year (2000), a total of

---

132 FCCC Subsidiary Body for Implementation (1997), *First Compilation and Synthesis of Second National Communications from Annex I Parties*, FCCC/SBI/1997/19, para. 7.
133 FCCC Secretariat (1998), *Second Compilation and Synthesis of Second National Communications from Annex I Parties: Summary*, FCCC/CP/1998/11, para. 4. Many states were also late with their third national communication: FCCC Subsidiary Body for Implementation (2003), *Compilation and Synthesis of Third National Communications (Annex I)*, para. 1.
134 A non-Annex I party was required to make its first national communication within three years of the entry into force of the FCCC for that party, or at such time as financial resources for the task were made available to it.
135 FCCC Subsidiary Body for Implementation (1999), *First Compilation and Synthesis of Initial National Communications from Non-Annex I Parties*, FCCC/SBI/1999/11, para. 1.
136 Ibid., para. 19.

27 non-Annex I parties had submitted their first national communications. All but South Korea had received some external financial support to prepare them.[137] Several parties reported on mitigation projects they had identified and needed funding for. These communications also included requests for financial assistance on adaptation.[138]

It is certainly possible that developing countries kept up with their reporting obligations under the climate change regime because they saw it as an opportunity to 'apply' for funding. However, this could only be a partial explanation, for it does not account for the thoroughness of most of the national communications and the increasing frequency of reporting in recent years. By 2001, the number of non-Annex I national communications for the first round had grown to 52. While the Secretariat found many positive things to say about them, the reporting on policies and measures remained vague and input-focused (a problem also vitiating Annex I communications): 'In many instances ... due to the limited information provided by Parties, it was extremely difficult to discern the level of implementation of the reported measures.'[139] Several of the developing countries reported on mitigation measures (and not just on adaptation measures), but only in a negative way. They cited a lack of public and political support for the implementation of such measures,[140] presumably meaning to emphasize the legal/political point that responsibility for mitigation rested mainly with Annex I parties.

The sixth and final instalment of the compilation-and-synthesis report on non-Annex I parties' first national communications is from 2005. Some of the content of the communications highlighted by the Secretariat seems inaccurate. For example, China reported that since the 1980s, through various policies and measures, it had succeeded in supporting rapid economic development with a relatively low growth rate of energy consumption and associated emissions.[141] China's claim seems self-serving and misleading. Its reported emissions rose from 3.7 Gt CO_2 eq. in 1994 to 7.0 Gt CO_2 eq. in 2005.[142] Improved reporting

---

137 FCCC Subsidiary Body for Implementation (2000), *Second Compilation and Synthesis of Initial National Communications from Non-Annex I Parties*, FCCC/SBI/2000/15, para. 9; see also FCCC Subsidiary Body for Implementation (2002), *Fourth Compilation and Synthesis of Initial National Communications from Non-Annex I Parties: Executive Summary*, FCCC/SBI/2002/8, para. 59 (most parties acknowledged having received financial assistance from the Global Environment Facility and its implementing agencies and other bilateral or multilateral programmes).
138 FCCC Subsidiary Body for Implementation (2000), *Second Compilation and Synthesis of Initial National Communications (Non-Annex I)*, paras 17, 41.
139 FCCC Subsidiary Body for Implementation (2001), *Third Compilation and Synthesis of Initial National Communications from Non-Annex I Parties: Executive Summary*, FCCC/SBI/2001/14, para. 13.
140 FCCC Subsidiary Body for Implementation (2002), *Fourth Compilation and Synthesis of Initial National Communications (Non-Annex I)*, para. 61.
141 FCCC Subsidiary Body for Implementation (2005), *Sixth Compilation and Synthesis of Initial National Communications from Non-Annex I Parties; Addendum: Inventories of Anthropogenic Emissions by Sources and Removals by Sinks of Greenhouse Gases*, FCCC/SBI/2005/18/Add.2, para. 13.
142 See Table 4.4 in Chapter 4.

should make it more difficult for states to get away with political slogans. Even the relatively mild process of International Consultation and Analysis which begins in 2015, represents a vast improvement in the quality control of non-Annex I reports (as Vihma, quoted earlier, has suggested).

Going by the Secretariat's multi-instalment report on the first round of non-Annex I national communications, the gaps in reporting are large and a certain extent of fudging of data and subterfuge are present. However, reporting by developing countries was, finally, nearly universal and there was no resistance in principle, at least, to what I have been calling accountable reporting.

On the Annex I front, the in-depth reviews by Expert Review Teams found that the transparency and completeness of the fifth national communications of Annex I parties could be improved in several ways. The following is a typical comment from an in-depth review report:

> The ERT encourages Germany to undertake a number of improvements regarding the transparency and completeness of its reporting ... (a) Provide more detailed information on factors affecting its GHG emissions; (b) Report also on non-mandatory reporting elements, including summary tables of PaMs [policies and measures] by sector; information on the monitoring and evaluation of PaMs; a quantification of emission reduction effects for all sectors; and a description of the PaMs influencing GHG emissions from international transport and activities regarding forestry; (c) Further elaborate on its methodologies and approaches for impact assessment, vulnerability and adaptation.[143]

It will be noted that these improvements are 'encouraged', not mandated. In some cases, national communications were missing sections, and information was internally inconsistent. The review of Australia's fifth national communication found that it did not include information required by the FCCC reporting guidelines explaining how financial support to developing countries was determined to be 'new and additional'.[144] Australia had claimed transfers of new and additional financial resources amounting to A$476 million for the period 2005–2009, including A$30 million for the GEF. Evidence that the claimed transfers were new and additional was simply left out of the national communication.[145] During the ERT review, Australia made a written submission stating that the financial support it had provided was 'new and additional' because it was financed from an increasing Official Development Assistance budget.[146] The ERT did not find the explanation adequate. It was unclear, it said, how the increased ODA was for the benefit of FCCC obligations, specifically. The ERT concluded that while Australia

---

143 Expert Review Team (2011), *Report of the in-Depth Review of the Fifth National Communication of Germany*, FCCC/IDR.5/DEU, para. 127.
144 Expert Review Team (2012), *Report of the in-Depth Review of the Fifth National Communication of Australia*, FCCC/IDR.5/AUS, p. 4.
145 Ibid., p. 31.
146 Ibid., p. 31.

included in its communication most of the mandatory information required by the FCCC guidelines, some mandatory reporting elements were missing and some areas of transparency could be improved. Prodded by the ERT during the review, Australia provided 'sufficient' additional information to make up for most of the missing mandatory reporting elements.[147]

While all in-depth reviews reported problems,[148] no review suggested that any country was falling behind the rest in steadily improving its reporting practices.

It should be noted that ERTs are not truly independent or objective actors in the FCCC's system of accountable reporting. As Fransen observes:

> Individuals involved in the review process [in-depth reviews of national communications] have commented in not-for-attribution interviews that the process is subject to weaknesses: parties at times pressure the review teams to alter the language used in the reports, the Subsidiary Body for Implementation does not consider individual reports, and parties are reluctant to challenge each others' communications for fear of their own communications being challenged.[149]

The weaknesses in the climate change regime's review system on state reporting become particularly evident under the light of the Kyoto Protocol's compliance system—a topic I take up in Chapter 3. However, even accepting the compromised nature of FCCC review, i.e. that it is less than perfect, the evidence summarized here suggests that there exists a significant level of state compliance with reporting obligations on national communications.

---

147 Ibid., p. 37.
148 The ERTs' most common recommendations for improvement in national communications included the following: More detail on the domestic climate change governance structure and the 'national system'. Stricter adherence to the FCCC reporting guidelines when reporting on policies and measures. Improved reporting, including quantitative estimates, on the rate of the expected total effect of implemented and adopted policies and measures on climate change. Information was lacking on the costs and cost-effectiveness of policies and measures. There was lack of clarity on which domestic institutions were responsible for monitoring, reporting, and evaluating policies and measures. More details were to be provided on the methodologies used for greenhouse gas emission projections. The financial assistance provided to developing countries was unclear, in particular how the test of 'new and additional' had been satisfied. More details were needed on the transfer of technology to developing countries. A clearer distinction was necessary between activities undertaken by the public sector and those undertaken by the private sector on technology transfer. Lastly, there was a lack of evidence on how the use of Kyoto Protocol mechanisms was supplemental to domestic action.
149 Taryn Fransen (2009), *Role of National Communications and Inventories*, p. 8. I return to this point in Section 3.2.1.

2.3.2 Assessment of inventories

Annex I parties generally complied with their obligation to submit greenhouse gas inventories annually. What still needs to be considered is the quality of the submitted inventories. Compared with the early years (the 1990s), state emission inventories have improved significantly. In 1994, the FCCC Secretariat reported that the Annex I inventories accompanying the national communications contained information gaps, which most often amounted to missing background data or inadequately documented methods. The minimum documentation standards for transparency were not always followed.[150] In 1996, the Secretariat attributed problems of transparency to a lack of experience in preparing inventory data as well as 'imperfections' in the guidelines. It conceded that the ability of Annex I parties to report emissions from the LULUCF sector was underdeveloped.[151] The quality of the inventory data submitted with the second national communications was higher than in the first round.[152] Yet, in 1998 the Secretariat observed that 'few Parties fully complied with the guidelines, in particular reporting data in tabular format'.[153] Compliance with reporting obligations by Annex I parties was not instant, but was to improve in steps over the course of two decades (1994–2014).

The reporting of emissions by developing countries has trailed so far behind that by developed countries as to be almost beyond comparison. In a compilation-and-synthesis report from 2000, the Secretariat noted that the two primary factors that affected the quality of non-Annex I emission inventories were the availability and quality of activity data and the updating of inventory data on a continuous basis by stable national teams.[154] Many countries simply did not further track their emissions once they had finished preparing a national communication. They complained that they did not have sufficient resources to do so.[155] By 2005, only 12 developing countries had reported their emissions for 1990, another 94 (77 per cent of non-Annex I parties) had reported their emissions for 1994, with the remainder providing an inventory for another year of their choosing.[156] Most developing countries conceded that their technical and institutional capacities were inadequate for meeting their reporting obligations under the FCCC regarding

---

150 FCCC Interim Secretariat (1994), *First Compilation and Synthesis of First National Communications (Annex I)*, para. 4.
151 FCCC Secretariat (1996), *Second Compilation and Synthesis of First National Communications (Annex I)*, para. 21.
152 FCCC Subsidiary Body for Implementation (1997), *First Compilation and Synthesis of Second National Communications (Annex I)*, para. 7.
153 FCCC Secretariat (1998), *Second Compilation and Synthesis of Second National Communications (Annex I)*, para. 4.
154 FCCC Subsidiary Body for Implementation (2000), *Second Compilation and Synthesis of Initial National Communications (Non-Annex I)*, para. 8.
155 FCCC Subsidiary Body for Implementation (2002), *Fourth Compilation and Synthesis of Initial National Communications (Non-Annex I)*, para. 60.
156 FCCC Subsidiary Body for Implementation (2005), *Sixth Compilation and Synthesis of Initial National Communications (Non-Annex I): Addendum*, para. 33.

their national inventories.¹⁵⁷ Thus even today a huge gap exists in the regularity and quality of emission reporting between the two groups.¹⁵⁸

The greenhouse gas inventories of Annex I parties are reviewed by Expert Review Teams, and this of course greatly benefits them, both because ERTs identify areas for improvement and because states try harder to meet the reporting requirements in order to minimize ERT criticism. However, it should not be thought that the ERT review process for Annex I inventories is thorough and uncompromising or that it ensures that at least these inventories are an entirely accurate representation of emissions from Annex I parties. To say that Annex I parties comply with their reporting obligations under the climate change regime is not to say that our grasp of how each of these states contributes to climate change is flawless. ERTs do occasionally raise serious non-compliance issues (see Chapter 3), but as a general rule they find that developed countries adequately account for their emissions. Should we trust the ERTs' judgement? Or, to put it in 'compliance' terms, does the FCCC's review system ensure that emission inventories, at least for Annex I parties, are transparent, complete, and accurate within the framework of the FCCC guidelines?

An Expert Review Team has at its disposal certain tests with which to gauge the accuracy of a state's accounting of its greenhouse gas emissions. The ERT will check that the standard IPCC methodologies have been followed by the state to calculate its emissions from each of the IPCC's emission sectors and subsectors.¹⁵⁹

An ERT is also provided with a reviewed state's inventories from earlier years, in most cases going back to 1990. A significant trend disruption from one year to the next will alert the ERT to an anomaly.¹⁶⁰ If the state has not provided an explanation for the anomaly in its annual submission accompanying its greenhouse gas inventory, the ERT may demand an explanation.

Another indicator that an ERT may rely on is country-level fuel-production and fuel-import/export statistics, available from the International Energy Agency.¹⁶¹ The IEA data serves as a reference line against which to plot the reviewed state's own estimates of emissions from its energy sector.¹⁶² The Food and Agriculture Organization is another source of country-level data that may be relied on by an ERT for the purpose of verifying state-reported emission inventories. Relevant factual and technical information from 'competent' NGOs may also be considered.¹⁶³

---

157 Ibid., para. 86. See also Table 4.4 in Chapter 4.
158 Compare Table 4.3 with Table 4.4 in Chapter 4.
159 Kyoto Protocol (2005), *Decision 22/CMP.1*, para. 65.
160 Ibid., para. 65.
161 Ibid., Annex, para. 65 ('The expert review team shall ... Compare the activity data of the Party included in Annex I with relevant external authoritative sources, if feasible, and identify sources where there are significant differences').
162 The IEA provides the FCCC Secretariat annually with datasets, reported by country, that include energy balances and net calorific values. The 'reference' and 'sectoral' approaches and their differences are discussed in, for example, International Energy Agency (2010), CO_2 *Emissions from Fuel Combustion: Highlights*, pp. 27–30.
163 Kyoto Protocol (2005), *Decision 27/CMP.1, Procedures and Mechanisms Relating to Compliance under the Kyoto Protocol*, FCCC/KP/CMP/2005/8/Add.3, Annex,

This is a rather limited set of tests. First, actual emissions are not normally measured by states directly but are estimated based on activity data and emission factors. As Fransen puts it, 'The expert review process for national communications falls short of verification, in that it assesses the document's adherence to reporting guidelines, rather than the reliability of reported information.'[164] Ulfstein and Werksman echo this point, observing that, 'Assessments made by an ERT ... will not be on the basis of end-of-pipe measurements, but rather will be based on whether a party had followed good practice in applying formulae that estimate and extrapolate emissions from input and output data.'[165]

Second, even in those Annex I states possessing relatively advanced inventory-generation systems, the uncertainties built into their greenhouse gas inventories are considerable. Two types of uncertainty account for this phenomenon.

The first could be called 'physical-technological uncertainty'. It comprises inaccuracies resulting from 'hard' uncertainties (concerning activity data and emission factors) in the calculation of the production and release of a given greenhouse gas from a given source.[166] Physical-technological uncertainty arises from the limitations of measuring instruments and our current knowledge of the underlying physical processes. The complexity of modelling highly variable sources of emissions over space and time, particularly for some biological sources, contributes to this type of uncertainty. For example, within the energy sector, emissions from transport can be estimated with much greater certainty than fugitive emissions from gas production and distribution. Uncertainty levels also vary across the IPCC sectors; e.g. overall emissions from the energy sector are known with much greater confidence than overall emissions from LULUCF.

The second type of uncertainty is caused by 'soft' systems instituted at the state level to gather and report data about greenhouse gas emissions inside the jurisdiction of each state. It leads to what I call 'regulatory-administrative uncertainty'. The cause of the uncertainty might be faulty or incomplete or biased reporting systems, or dishonestly implemented systems; these will tend to capture fewer data on greenhouse gas emissions than are actually available (or are straightforwardly available) at the physical-technological level, and thus will tend to under-report emissions even of gases that are relatively easy to measure. For example, it is common for states to consider certain greenhouse gas emissions as 'negligible' and not to report them for this reason. Or a state's authorities may consider that they have no practical way of gaining information about certain emission sources. A budget cut to a state's statistics bureau might cause it to cease to collect certain types of information. Moreover, a state may choose not to report

---

para VIII.4.
164 Taryn Fransen (2009), *Role of National Communications and Inventories*, p. 12.
165 Geir Ulfstein and Jacob Werksman (2005), 'The Kyoto Compliance System: Towards Hard Enforcement', in *Implementing the Climate Regime: International Compliance*, ed. Jon Hovi, Olav Stokke, and Geir Ulfstein, p. 52.
166 The same of course is true of the absorption of greenhouse gases by sinks.

emission information it considers confidential for commercial or national-security reasons.[167]

Another cause of gaps in reporting is the fact that the IPCC has not developed a methodology for the calculation of every source of anthropogenic greenhouse gas emissions in a state,[168] and where no such methodology exists the state need not report emissions from that source.[169] It had been anticipated that domestic NGOs would act as watchdogs and maintain a bottom-up pressure on governments to improve their collection of greenhouse gas emission data, but such a pressure has not materialized.[170]

---

167 FCCC, art. 12(9), broadly provides for confidentiality where requested by a party, subject to criteria to be developed by the COP.
168 Terje Berntsen, Jan Fuglestvedt, and Frode Stordal (2005), 'Reporting and Verification of Emissions and Removals of Greenhouse Gases', in *Implementing the Climate Regime: International Compliance*, ed. Jon Hovi, Olav Stokke, and Geir Ulfstein, p. 86.
169 FCCC Secretariat (undated), *Handbook for Review of National GHG Inventories*, ch. 2, p. 13, leaves any follow-up to an ERT's discretion: 'The UNFCC reporting guidelines encourage Parties to estimate all existing (anthropogenic) source and sink categories, including sources/sinks for which there are no agreed IPCC methodologies. However, it may be inappropriate to expect a particular Party to provide an estimate of a country-specific source/sink when estimating such a source/sink would divert resources from key categories, unless that source/sink is likely to be significant. The ERT should, therefore, consider the likely significance of an unreported country-specific source, as well as the overall key categories of the Party, in evaluating whether to encourage the Party to investigate the significance of the source.'
170 State non-compliance with routine obligations of the climate change regime has not been a theme in the programmes of major environmental, social-justice, and human-rights NGOs. Their reports certainly suggest that states must to be prodded to act on environmental protection so as to prevent them giving in to the demands of lobby groups with opposing interests, such as 'Big Oil' or coal. But the main fight for most NGOs is not against states but against other pressure groups. Their focus is on concrete projects undertaken in both public and private partnerships. Their main effort is to increase the number of beneficial projects on the ground and block counterproductive legislative bills or industrial expansion, on the theory that eventually the tide will turn. NGOs generally engage with issues that the public can relate to and can generate qualified optimism about the future. Whaling and trade in endangered species, for example, are moreover easier for NGOs to monitor than reporting on atmospheric chemistry (see Jørgen Wettestad (2007), 'Monitoring and Verification', in *The Oxford Handbook of International Environmental Law*, ed. Daniel Bodansky, Jutta Brunnée, and Ellen Hey, p. 986). Thus NGOs do not, in practice, act as watchdogs for accountable reporting of greenhouse gas emissions by states. For illustrations of the types of campaign actually engaged in, see Environmental Defense Fund (2013), *Annual Report 2013*, p. 34 (research projects to measure fugitive emissions from shale-gas wells, pipelines, and storage and distribution facilities, which is perhaps the closest an NGO has come to the subject of accountable reporting); and ibid., p. 38 (assistance given to China to launch carbon-trading pilot programmes); Sierra Club Foundation (2012), *Annual Report 2012*, pp. 10–11 (construction of a utility-scale solar plant); Greenpeace International (2012), *Annual Report 2012*, p. 13 (campaign to oppose the expansion of coal mining in biodiversity-rich forests); and Friends of the Earth International (2012), *Annual Report 2012*, p. 5 (campaign against influence of fossil-fuel corporations on UN policymaking).

In summary, and notwithstanding the various validity checks available to an Expert Review Team during a review of a state's greenhouse gas inventory, an ERT does not have access to independent information that would allow it to check the inventory's completeness and reliability. The tests either involve checking compliance with procedural (mainly IPCC) guidelines,[171] or are consistency checks. It has been said that 'in most cases the most important sources are well known, and scientific uncertainties occur mostly in relatively less important sources. ... [T]he vast majority of the greenhouse gas emissions in the Kyoto Parties is due to carbon dioxide from fossil fuel combustion',[172] and this information is generally accurate. However, as I discuss in the next section, the uncertainties are high enough to raise questions about the sufficiency of the FCCC's reporting-and-review system as a reliable guide to policy.

2.3.3 Independent scientific verification

In the state inventories of Annex I parties, uncertainty in the estimation of emissions from deforestation, reforestation, and forest degradation is relatively high, ranging from 25 to 100 per cent. Uncertainty concerning agricultural emissions is of the same order. When all uncertainty is combined, the overall amount is about the same as the expected emission reductions from the Kyoto Protocol's first commitment period (around 5 per cent below 1990 levels).[173]

As we have seen, accountable reporting of greenhouse gas emissions under the international climate change regime is limited to a minority group of countries including most major economies but not all of them. Only Annex I parties prepare inventories annually, and only Annex I inventories are independently checked. Shifting economic trends have led to a situation where, since 2005, most greenhouse gas emissions from human activities originate in developing countries.[174] The

---

171 See IPCC (1996), *Revised Guidelines*; IPCC (2000), *Good Practice Guidance*; and IPCC (2003), *Good Practice Guidance for LULUCF*.
172 Rob Swart et al. (2007), 'Are National Greenhouse Gas Emissions Reports Scientifically Valid?', 7 *Climate Policy* 535, p. 537.
173 Wilfried Winiwarter (2007), 'National Greenhouse Gas Inventories: Understanding Uncertainties Versus Potential for Improving Reliability', 7 (4–5) *Water, Air, and Soil Pollution: Focus* 443, p. 17 ('With trend uncertainties of several percentage points being typical of industrialized countries, reduction targets of 6–8% as formulated in the Kyoto protocol cannot be monitored unambiguously'); and, for comments to the same effect, see US National Research Council (2010), *Verifying Greenhouse Gas Emissions: Methods to Support International Climate Agreements*, pp. 4, 28. On the global uncertainty levels for various gases, see IPCC (2014), *Climate Change 2014: Mitigation of Climate Change: Working Group III Contribution to the Fifth Assessment Report of the Intergovernmental Panel on Climate Change: Summary for Policymakers*, p. 7 ('Global CO_2 emissions from fossil fuel combustion are known within 8% uncertainty (90% confidence interval). CO_2 emissions from FOLU [forestry and other land use] have very large uncertainties attached in the order of ±50%. Uncertainty for global emissions of CH_4, N_2O and the F-gases has been estimated as 20%, 60% and 20%, respectively').
174 Using fuel-production and import data and other general indicators it is possible to estimate that non-Annex I countries accounted for over 50 per cent of global GHG

largest and growing proportion of global anthropogenic emissions is thus either not regularly reported or not independently reviewed.[175] With the exception of the Annex I group, our knowledge of state responsibility for anthropogenic emissions since 1990 is significantly uncertain, and will remain so for the foreseeable future.

Improvements to our understanding of global emissions could come from making reporting of emissions more frequent (as has happened in the case of non-Annex I national communications and biennial update reports for the 2013–2020 period) and extending the independent review procedures to cover more countries (as is being done through the mild ICA process, covering the same period). Improvements could also arise from encouraging countries to use higher methodological 'tiers' (IPCC methodologies of higher sophistication) in their reporting. Constant expansion and technical refinement of emission reporting is certainly necessary if there is to be a global budgetary approach to mitigation with specific allocations to states (see Chapter 4). Improved reporting is to some extent driven by the IPCC's on-going work on methodological coverage, but for the most part reliance must be placed on the willingness of states to keep improving the FCCC's system of reporting and review. Is there an alternative to these piecemeal, gradualist efforts to make the current system more reliable?

A scientific specialization has recently formed that promises to offer just such an alternative. The group aspires to have state reporting under the FCCC checked by direct atmospheric measurements from ground stations and satellites. The advantages of direct verification are that it would measure sources directly and be independent of the subjective judgements about sources that are an unavoidable element of the preparation of a state's greenhouse gas inventory. 'The atmosphere does not misrepresent data or make mistakes; nor does it bend to ideology or political will.'[176] If state-reported emissions (bottom-up estimates) could be reconciled with those calculated from atmospheric measurements (top-down), there would be greater certainty about the amount of anthropogenic emissions and possibly even of state-level attribution.

The number of scientists who specialize in measuring greenhouse gases in the atmosphere is still small (around two hundred worldwide, by one estimate).[177] They have constructed a global network of CO_2-monitoring stations, many of which also measure the concentration of methane and other greenhouse gases.[178]

---

emissions in 2005, and are expected to account for an increasing proportion in the future: Jane Ellis, Sara Moarif, and Greg Briner (2010), *Core Elements of National Reports*, COM/ENV/EPOC/IEA/SLT(2010)1, p. 10. See also Glen P. Peters et al. (2012), 'Rapid Growth in CO_2 Emissions after the 2008–2009 Global Financial Crisis', 2 *Nature Climate Change* 2, p. 3.

175 U.S. National Research Council (2010), *Verifying Greenhouse Gas Emissions*, p. 29.
176 Editorial (9 February 2012), 'Gas and Air', 482 *Nature* 131, p. 132.
177 Andrew C. Manning et al. (2011), 'Greenhouse Gases in the Earth System: Setting the Agenda to 2030', 369 *Phil. Trans. R. Soc. A* 1885, p. 1885. The top-down approach to measuring greenhouse gases is also discussed in other articles in this special issue of the *Philosophical Transactions*.
178 Ibid., p. 1885. Data from the individual monitoring stations may be obtained from the World Meteorological Organization, 'World Data Centre for Greenhouse Gases',

The network's small size (only about 120 monitoring stations) means that there are large gaps for regions such as Africa, Asia, and South America—places distant from the main research centres of atmospheric scientists.[179] However, readings at the CO_2-monitoring stations are being supplemented by satellite remote sensing. A satellite provides a more general view than a ground station, and its data can be checked against the highly accurate measurements on the ground.[180] A NASA satellite called Orbiting Carbon Observatory-2 dedicated to the measurement of greenhouse gases from space went into orbit on 2 July 2014.[181]

The third element in this developing field of knowledge consists of modelling tools (trajectory models) that combine information about airborne greenhouse gases measured at multiple ground stations or from satellites, with information on wind flows. The models allow scientists to work backward from measured emissions to the sources of those emissions, and thus locate emission sources in the territory of a particular state.[182] 'Inverse modelling' is vital for state attribution of emissions and could serve the independent assessment of state inventories.

Testing of the new methods of direct scientific measurement is being reported with greater frequency. For example, in 2011, scientists from the US National Oceanic and Atmospheric Administration and the University of Colorado at Boulder studied methane emissions from a Denver natural-gas operation where 'fracking' methods were used. The scientists estimated cumulative fugitive emissions from the gas field using concentrations of pollutants in air samples. They found that the industry's reports underestimated the methane emissions by about 50 per cent.[183]

The growing scientific interest in the possibility of top-down verification of state emissions has received some institutional support, although as yet little from state parties to the FCCC. In 2010, the US National Research Council through its Committee on Methods for Estimating Greenhouse Gas Emissions, released a report on the quality of state inventories. It asked what could be done to lower the uncertainties. The short-term potential of top-down scientific initiatives was described in the report as follows:

<http://ds.data.jma.go.jp/gmd/wdcgg/>.
179 Andrew C. Manning et al. (2011), 'Setting the Agenda' p. 1886.
180 Ibid., pp. 1886–1887.
181 NASA Jet Propulsion Laboratory, 'OCO-2: Orbiting Carbon Observatory', <http://oco.jpl.nasa.gov/>. The first OCO satellite was destroyed during launch. OCO-2 is able to make around 100,000 measurements of CO_2 around the world every day and can pinpoint emissions on scales as small as a city building or street; see 'NASA's OCO-2 Will Track Our Impact on Airborne Carbon', <http://www.jpl.nasa.gov/news/news.php?release=2014-204>.
182 Andrew C. Manning et al. (2011), 'Setting the Agenda', p. 1888.
183 Editorial (2012), 'Gas and Air', p. 132. See also Mykola Gusti and Matthias Jonas (2010), 'Terrestrial Full Carbon Account for Russia: Revised Uncertainty Estimates and Their Role in a Bottom-up/Top-Down Accounting Exercise', 103 *Climatic Change* 159.

> Strategic investments would, within 5 years, improve reporting of emissions by countries and yield a useful capability for independent verification of greenhouse gas emissions reported by countries. ... fossil-fuel CO_2 emissions could be estimated by each country and checked using independent information with less than 10 per cent uncertainty. The same is true for satellite-based estimates of deforestation, which is the largest source of CO_2 emissions next to fossil-fuel use, and for afforestation, which is an important sink for CO_2. However, self-reported estimates of N_2O, CH_4, CFC, HFC, PFC, and SF_6 emissions will continue to be relatively uncertain and we will have only a limited ability to check them with independent information.[184]

The 'strategic investments' would be for the purpose of putting in place more equipment for atmospheric monitoring in ground-based stations and in satellites. While such a programme would be expensive,[185] it could save governments money in the long run.[186] New measurement stations would have to be established near cities and other large sources of greenhouse gas emissions,[187] as well as in developing countries, where few exist. More CO_2-sensing and forest-cover-reading satellites would have to be launched to strengthen current capacity; improvements in the measurement of radiocarbon (^{14}C) would be necessary to enable fossil-fuel CO_2 emissions to be distinguished from CO_2 from other sources;[188] tracer-transport inversion models (the trajectory models mentioned above) would have to be further developed, to improve their predictive capacity;[189] and the bottom-up inventories produced by states would have to be gridded at spatial and temporal resolutions to assist with independent verification using the top-down method.[190]

The scientific developments summarized in this section suggest that an independent, top-down, check on the emissions reported by states is possible in the not-too-distant future. Under an enhanced review or compliance regime,

---

[184] US National Research Council (2010), *Verifying Greenhouse Gas Emissions*, p. 1. See also Rob Swart et al. (2007), 'Are National Greenhouse Gas Emissions Reports Scientifically Valid?', pp. 537–538 ('Further steps toward stabilization will require larger emissions reductions and hence lead to stronger atmospheric signals, which in the case of methane and other non-CO_2 greenhouse gases would be easier to verify with inverse modeling').

[185] Establishing 5–10 new ground measurement stations per year in the United States would require a budget of $15–20 million per year. The satellite OCO-2, with a two-year lifetime, cost around $278 million. US National Research Council (2010), *Verifying Greenhouse Gas Emissions*, pp. 63–64.

[186] A. J. Durant et al. (2011), 'Economic Value of Improved Quantification in Global Sources and Sinks of Carbon Dioxide', 369 *Phil. Trans. R. Soc. A* 1967.

[187] US National Research Council (2010), *Verifying Greenhouse Gas Emissions*, pp. 5–6.

[188] Ibid., pp. 6–7, 61f.

[189] Ibid., pp. 53f. See also P. Ciais et al. (2010), 'Atmospheric Inversions for Estimating CO_2 Fluxes: Methods and Perspectives', 103 *Climatic Change* 69; and L. Rivier et al. (2010), 'European CO_2 Fluxes from Atmospheric Inversions Using Regional and Global Transport Models', 103 *Climatic Change* 93.

[190] US National Research Council (2010), *Verifying Greenhouse Gas Emissions*, pp. 34–35.

National reporting and assessment 61

procedural verification by an international treaty body, namely the FCCC's Expert Review Teams, could be supplemented by independent scientific checks. Of course, the international climate change regime might never agree to independent verification of state reports by scientists. Even if it did, it would take years for the two mechanisms to mesh legally and in practice. Thus independent scientific verification is unlikely to have any impact on FCCC reporting rules and compliance monitoring in the decade to 2020.

2.4 Conclusion: 'Accountable reporting' as a driver of compliance with substantive obligations

The requirement of article 4 of the FCCC for a periodic production of national communications is an exercise that requires states to articulate and streamline their response to the climate change problem, identify gaps in their actions, and compare and contrast their performance with that of other states.

Bodansky writes that 'national reporting can perform a policy reform function by encouraging self-examination. ... the process of preparing a national report may have a catalytic effect in promoting internal policy reform by mobilizing and empowering actors both within and outside the government'.[191] Faure and Lefevere have suggested that better reporting leads to better compliance with substantive obligations: 'The development of a more elaborate and transparent system for the provision of information on the compliance of ... states automatically increases [their] accountability.'[192] An explanation of how this works is usually given in quasi-psychological terms:

> Transparency ... reassures parties that others are meeting their obligations; and if they are not, it permits a timely response. Reassurance is needed when actors otherwise inclined to comply are concerned that they will be placed at a disadvantage if their compliance is not matched by others.[193]

The advent of IAR and ICA could be taken to represent what Mehling calls 'a general shift towards processes of assessment, deliberation, and justification'

---

191 Daniel Bodansky (2010), *The Art and Craft of International Environmental Law*, p. 239.
192 Michael G. Faure and Jürgen Lefevere (2011), 'Compliance with Global Environmental Policy', in *The Global Environment: Institutions, Law, and Policy*, ed. Regina S. Axelrod, Stacy D. VanDeveer, and David Leonard Downie, p. 179.
193 Abram Chayes, Antonia Handler Chayes, and Ronald B. Mitchell (1998), 'Managing Compliance: A Comparative Perspective', in *Engaging Countries: Strengthening Compliance with International Environmental Accords*, ed. Edith Brown Weiss and Harold K. Jacobson, p. 44. See also William Hare et al. (2010), 'The Architecture of the Global Climate Regime: A Top-Down Perspective', 10 (6) *Climate Policy* 600, pp. 607-8 ('International review/verification is necessary in order to build trust between countries ... Without a centralized system for the monitoring, review and verification of emission reductions, only those countries with the capacity to undertake such activities unilaterally would be able to detect whether others were fulfilling their obligations').

62 National reporting and assessment

as opposed to a 'narrow definition of enforcement as punishment or coercion'.[194] Rajamani recognizes that the IAR and ICA processes 'are intended ... to perform a compliance role in the future climate regime.'[195] The question is: How far into the future?

The US view on the post-2020 agreement implies that the FCCC's accountable reporting system for the 2013-2020 period is sufficient as a compliance system to supervise all important treaty obligations, and that a separate compliance system with a compliance committee and punitive powers, such as that found in the Kyoto Protocol, is not necessary for the obligations under the new agreement:

> each Party should be required to report periodically on its progress in implementing its schedule [i.e. the document reflecting the party's pledged contribution to the global effort to limit/reduce GHG emissions]. Reporting will enhance ambition by promoting accountability, by making countries take a hard look at their inventories and mitigation opportunities, by revealing best practices, and by enabling assessment of the aggregate global effort. [A]ll Parties should be expected to follow the same set of agreed guidelines, recognizing that such guidelines should provide for appropriate differentiation in light of capabilities and circumstances.[196]

Moreover, the United States has suggested that some form of the IAR/ICA system should be used for reviewing state reports and naming and shaming states for underperformance:

> the agreement should provide for Parties' implementation of their schedules to be reviewed. Such reviews ... will enable other Parties and the international community to know the extent to which schedules are being implemented; they will allow the Parties to assess the sufficiency of the global effort in the aggregate; and they will provide a powerful incentive for Parties to engage in meaningful implementation of their schedules. Reviews should be based on a single system.'[197]

What the United States seems to be saying is that while accountable reporting is a good thing, there is no reason to take it to extremes.

In conclusion, an examination of state practice reveals that state reporting and review represent a well-established rule of the international climate change regime. Accountable reporting, at least in the form it has acquired under the FCCC for the 2013–2020 period, is so fundamental to the thinking of states as to qualify as climate law.

---

194 Michael Mehling (2012), 'Enforcing Compliance in an Evolving Climate Regime', in *Promoting Compliance in an Evolving Climate Regime*, ed. Jutta Brunnée, Meinhard Doelle, and Lavanya Rajamani, p. 196.
195 Lavanya Rajamani (2012), 'Developing Countries and Compliance', p. 382.
196 United States (2014), *US Submission on Elements of the 2015 Agreement*, p. 5.
197 Ibid., pp. 5-6.

3 Facilitation and enforcement of rules through the Kyoto Protocol's Compliance Committee

Reporting and review requirements under the FCCC and the Kyoto Protocol (national communications, greenhouse gas inventories, and in-depth, technical, periodic, and annual reviews of the information submitted by states) were discussed in Chapter 2. For Annex I parties which are also parties to the Kyoto Protocol there remains a group of rules at the tail end of the review process that is supervised by a distinct and unique institution: the Compliance Committee. In the climate change regime, the notion of compliance is not exclusive to the Protocol (states must, of course, comply with their FCCC obligations). What the Kyoto Protocol introduced was the notion of compliance through 'enforcement' by a specialized body having the right to impose penalties. These elements raise special questions of compliance, including whether a Protocol-style compliance system is a necessary element of the law on accountable reporting.

3.1 Rules and process

The sections that follow review the rules that apply to Annex I parties to the Kyoto Protocol in relation to 'questions of implementation' and the functions of the Compliance Committee and its two branches. In the second part of the chapter I assess the legacy of the Protocol's compliance system.

3.1.1 Questions of implementation

As discussed in Chapter 2, an Expert Review Team is to provide a thorough and comprehensive technical assessment of all aspects of a party's implementation of the Kyoto Protocol 'and identify any problems in, and factors influencing, the fulfillment of commitments'.[1] If the ERT identifies a 'potential problem' during a review, it must question the party about the problem and offer advice

---

1 Kyoto Protocol (2005), *Decision 22/CMP.1, Guidelines for Review under Article 8 of the Kyoto Protocol*, FCCC/KP/CMP/2005/8/Add.3, Annex, para. 4.

on how it could be corrected.[2] A potential problem is a precursor to a question of implementation, which is a *confirmed* problem.[3]

The work of Expert Review Teams is guided by an outlook that can only be described as facilitative. The ERTs' facilitative role is emphasized by the compliance system's objectives: ERTs are to 'promote consistency and transparency in the review of information' submitted by Annex I parties to the Protocol; and they are to 'assist' these parties to improve their reporting of required information and, in general, their commitments under the Protocol.[4] An ERT's official advisory role further affirms the body's facilitative outlook: 'The expert review team should offer advice to Parties included in Annex I on how to correct problems that they identify, taking into account the national circumstances of the Party'.[5]

Expert Review Teams potentially also have a more confrontational side, although in practice it is rarely seen. They can raise a potential problem and escalate it into a question of implementation. From a legal point of view, the latter is a prima facie finding that the state is not in compliance with a rule of the Protocol. ERT review reports are forwarded by the FCCC Secretariat to the Compliance Committee, and in this manner any question of implementation is brought to the Committee's attention.[6] Once the Committee receives a question of implementation, it initiates its own procedure to engage the allegedly non-compliant state.

A fundamental question arising in this context was adumbrated in Chapter 2: What types of non-compliance with the Protocol may ERTs list as questions of implementation? An answer is provided in a 2005 decision of the parties:

> if an unresolved problem *pertaining to language of a mandatory nature in these guidelines influencing the fulfilment of commitments* still exists after the Party included in Annex I has been provided with opportunities to correct the problem within the time frames established under the relevant review procedures ... that problem [shall] be listed as a question of implementation in the final review reports. An unresolved problem pertaining to language of a non-mandatory nature in these guidelines shall be noted in the final review report, but shall not be listed as a question of implementation.[7]

It is clear from this passage that an ERT must ('shall') list every instance of non-compliance with a mandatory requirement of the Protocol as a question of implementation. Any shortcoming in state practice that does not breach a

---

2 Ibid., para. 7.
3 Questions of implementation can also be raised by states themselves. An example of this will be given in Section 3.2.2, below.
4 Kyoto Protocol (2005), *Decision 22/CMP.1*, paras 2.b and 2.c.
5 Ibid., para. 5; and see, ibid., paras 106 and 117.
6 Kyoto Protocol (2005), *Decision 27/CMP.1, Procedures and Mechanisms Relating to Compliance under the Kyoto Protocol*, FCCC/KP/CMP/2005/8/Add.3, Annex, para. VI.1.
7 Kyoto Protocol (2005), *Decision 22/CMP.1*, Annex, para. 8.

mandatory rule must *not* be listed it as a question of implementation but must be mentioned as an unresolved problem in the ERT's review report.

The word 'mandatory' occurs only once in the Protocol's instructions on ERT reviews—at the point quoted above. A separate decision, setting out the procedures and mechanisms relating to compliance,[8] makes no reference at all to the concept. The italicized segment of the passage I have just quoted is poorly phrased for a key legal provision. What, for example, does 'pertaining' mean in the phrase 'problem pertaining to language' and how are we to understand 'language ... influencing the fulfilment of commitments'?

One possibility, considered in Chapter 2, is that 'language of a mandatory nature' means any directive prefaced by a prescriptive 'shall' and not a discretionary 'should' or 'may'. The connotation of 'mandatory elements', where the term is applied to the rules concerning national communications and greenhouse gas inventories, was a puzzle under the FCCC, as well. The issue there never came to a head, mainly because the FCCC has no compliance system to create a demand for clear and workable definitions. In the Protocol's case, the stakes are much higher.[9]

The Protocol's guidelines for review[10] do not themselves set out how parties must calculate their emissions, organize their national systems, run their registries, or report on these matters and others. The 'language of a mandatory nature in these guidelines' mostly concerns language in *other* guidelines, some developed under the FCCC and others under the Kyoto Protocol. For example, in the Protocol's review guidelines, under the section on the review of annual inventories, 'Problems should be identified as a failure to follow agreed guidelines [on] Transparency, as defined in the UNFCCC reporting guidelines on annual inventories'.[11] Under the section on the review of national systems, 'problems' are to be identified on the basis of what is required of parties in the (separate) guidelines on national systems.[12] These guidelines do not offer a straightforward way to distinguish mandatory from non-mandatory elements.[13] The same holds for the guidelines on national registries.[14] Under the section on the review of national communications,

---

8 Kyoto Protocol (2005), *Decision 27/CMP.1*.
9 Mitchell writes that where 'negative responses such as sanctioning of noncompliant states [are used] the regime's rules will need to distinguish clearly between desirable and undesirable behavior, that is, between compliance or noncompliance': Ronald B. Mitchell (1998), 'Sources of Transparency: Information Systems in International Regimes', 42 *International Studies Quarterly* 109, p. 114. In the ramping up of the FCCC's accountable reporting system to the enhanced level of the Protocol's compliance system, the need for greater definitional clarity was overlooked.
10 The Annex to Decision 22/CMP.1 is entitled 'Guidelines for Review Under Article 8 of the Kyoto Protocol'.
11 Kyoto Protocol (2005), *Decision 22/CMP.1*, Annex, para. 69.
12 Ibid., Annex, paras 102–105.
13 E.g. 'National systems should be designed and operated to ensure the quality of the inventory' (Kyoto Protocol (2005), *Decision 19/CMP.1, Guidelines for National Systems under Article 5, Paragraph 1, of the Kyoto Protocol*, FCCC/KP/CMP/2005/8/Add.3, Annex, para. 7), although clearly this is a mandatory requirement.
14 Kyoto Protocol (2005), *Decision 22/CMP.1*, Annex, paras 115–116.

the Protocol's review guidelines say only: 'The problems identified during the assessment relating to individual sections of the national communication ... shall be identified as relating to Transparency; Completeness; Timeliness'.[15] The result is that the meaning of 'mandatory nature' is far from clear.

The FCCC's reporting guidelines referenced by the Protocol's review guidelines take a casual approach to the use of 'should' and 'shall'. As late as 2013, the FCCC parties were approving guidelines whose mandatory and non-mandatory elements could not be told apart on the basis of linguistic or other straightforward indicators.[16]

As a consequence of these drafting weaknesses, an Expert Review Team, obliged by the Protocol's rules to list non-compliance with mandatory elements of the Protocol as questions of implementation, must pause to consider what is mandatory and what is not, and the answer cannot always be uncontroversial. ERTs are, therefore, forced to assume a degree of discretion in order to continue to discharge their functions.

3.1.2 Compliance Committee structure

The Kyoto Protocol's Compliance Committee is to 'facilitate, promote and enforce compliance with the commitments' of states under the Protocol. The Committee is constituted of a Facilitative Branch, an Enforcement Branch, and a Bureau. The Bureau must decide whether to assign a question of implementation raised by an Expert Review Team to the Facilitative Branch or the Enforcement Branch.[17]

The Bureau's assignment of a question of implementation to one of the branches must be 'in accordance with the mandates of each branch'.[18] For reasons that will become clear later, it is important to mention the fact that in the Protocol decision

---

15 Ibid., Annex, para. 137.
16 For example: 'Each Annex I Party should implement and maintain national inventory arrangements for the estimation of anthropogenic GHG emissions by sources and removals by sinks ... each Annex I Party should ... Designate a single national entity with overall responsibility for the national inventory ... each Annex I Party should ... Prepare national annual GHG inventories in a timely manner in accordance with these reporting guidelines and relevant decisions of the COP ... GHG emissions and removals should be presented on a gas-by-gas basis'; but: 'Annex I Parties shall report actual emissions of HFCs, PFCs, SF_6 and NF_3, providing disaggregated data by chemical ... Annex I Parties shall estimate and report the individual and cumulative percentage contributions from key categories to their national total', etc.: FCCC (2013), *Decision 24/CP.19, Revision of the UNFCCC Reporting Guidelines on Annual Inventories for Parties Included in Annex I to the Convention*, FCCC/CP/2013/10/Add.3, Annex.
17 Kyoto Protocol (2005), *Decision 27/CMP.1*, Annex, para. VI.1. The annex is entitled 'Procedures and Mechanisms Relating to Compliance Under the Kyoto Protocol'. Andresen and Gulbrandsen write that the idea of a dual approach to compliance was jointly introduced to the Kyoto Protocol negotiations by two NGOs, namely WWF and the Center for International Environmental Law: Steinar Andresen and Lars H. Gulbrandsen (2005), 'The Role of Green NGOs in Promoting Climate Compliance', in *Implementing the Climate Change Regime: International Compliance*, ed. Jon Hovi, p. 175.
18 Kyoto Protocol (2005), *Decision 27/CMP.1*, Annex, para. VII.1.

setting up the Compliance Committee there is no procedure for allocating any sort of problem to one of the branches other than a question of implementation. There is not even a mention of any other problem type. Likewise, the Committee's Rules of Procedure[19] do not envisage either branch engaging with any problem other than a question of implementation.

Members of the Compliance Committee and their alternates are to serve on the Committee in their individual capacity. They must have competence relating 'to climate change and in relevant fields such as the scientific, technical, socioeconomic or legal fields'.[20] All members of the Enforcement Branch are to have 'legal experience'.[21] This does not mean that they must be legally qualified. It has been suggested that the Protocol's Compliance Committee is a quasi-judicial construct.[22] While it has aspects of a legal procedure, by design it does not specifically require any lawyers to be assigned to the Committee, although in practice several of its members have held legal qualifications.[23]

Another common assumption is that the Compliance Committee has low discretion and high 'automaticity'.[24] This is true only of certain aspects of its design. Other operational aspects are not strictly defined, and a few of them explicitly bestow a discretion on Committee members. For example, the Committee 'shall take into account any degree of flexibility allowed by the [Conference of the Parties to the Protocol] to the Parties included in Annex I undergoing the process of transition to a market economy.'[25] This is a highly discretionary power. Moreover, the Facilitative Branch, which, as we shall see, was poorly conceived and designed in the first place, has in recent years claimed a discretion to redefine its role. I return to this matter below.

The Compliance Committee became operational in 2006.[26]

---

19 *Rules of Procedure of the Compliance Committee of the Kyoto Protocol*, contained in the annex to Decision 4/CMP.2, as amended by Decision 4/CMP.4 and further amended by Decision 8/CMP.9.
20 Kyoto Protocol (2005), *Decision 27/CMP.1*, Annex, para. II.6.
21 Ibid., Annex, para. V.3.
22 See, e.g., Sebastian Oberthür and René Lefeber (2010), 'Holding Countries to Account: The Kyoto Protocol's Compliance System Revisited after Four Years of Experience', 1 (1) *Climate Law* 133, p. 140.
23 On this topic, see Geir Ulfstein and Jacob Werksman (2005), 'The Kyoto Compliance System: Towards Hard Enforcement', in *Implementing the Climate Regime: International Compliance*, ed. Jon Hovi, Olav Stokke, and Geir Ulfstein, pp. 45, 47.
24 E.g. Sebastian Oberthür and René Lefeber (2010), 'Compliance System Revisited', p. 141; Jutta Brunnée (2012), 'Climate Change and Compliance and Enforcement Processes', in *International Law in the Era of Climate Change*, ed. Rosemary Rayfuse and Shirley V. Scott, p. 306.
25 Kyoto Protocol (2005), *Decision 27/CMP.1*, Annex, para. II.11. Another example: 'The facilitative branch shall [take] into account the principle of common but differentiated responsibilities and respective capabilities': ibid., Annex, para. IV.4.
26 Compliance Committee Plenary (29 May 2006), *Report on the First Meeting*, CC/1/2006/4.

3.1.3 Powers of the Facilitative Branch

The Facilitative Branch is to provide 'advice and facilitation' to the Kyoto Protocol parties in their implementation of the treaty's provisions. This relates to *all* parties. With respect to Annex I parties, which is the only group with 'commitments' under the Protocol, the Facilitative Branch is to 'promote' compliance with their emission-reduction, institutional, and reporting commitments.[27]

Within its overall mandate, the Facilitative Branch has a responsibility to address questions of implementation, but only insofar as they fall outside the mandate of the Enforcement Branch.[28] Phrased positively, a question of implementation for the Facilitative Branch may relate to article 3.14 of the Protocol (on the minimization of adverse social, environmental, and economic impacts of Protocol-related measures on developing parties), including how an Annex I party is 'striving' to implement that provision. A question of implementation may also be raised in relation to evidence that an Annex I party's use of the Protocol's flexibility mechanisms is not 'supplemental' to domestic action.[29]

The Facilitative Branch also has advisory/supportive and early-warning functions. It is not clear how these are meant to be engaged, but almost certainly it is not through the device of questions of implementation. The rules say only that the branch may seek to promote compliance and give early warning of potential non-compliance by providing 'advice and facilitation' to Annex I parties in relation to their commitments under article 3.1 of the Protocol (meeting quantified mitigation targets), article 5.1/5.2 (having a national system for the estimation of emissions and using approved methodologies to estimate emissions), and article 7.1/7.4 (reporting supplementary information in national communications and inventories). Matters relating to the last two may be addressed only prior to the beginning of the first commitment period, whereas in the case of the first, promotion of compliance and early warning of potential non-compliance may be pursued both prior to and for the duration of the relevant commitment period.[30] At other times, problems arising in connection with these matters will be questions of implementation that go to the Enforcement Branch.

The rules specify a set of permissible responses to states that the Facilitative Branch may choose from when undertaking its functions. The rules call these responses 'consequences'—a case of 'diplomatese' to avoid using the more accurate but guilt-connoting terms 'responses' and 'measures' (and, in the case of what the Enforcement Branch can apply, 'penalties'). Four types of response

---

27 Kyoto Protocol (2005), *Decision 27/CMP.1*, Annex, para. IV.4.
28 Ibid., Annex, para. IV.5.
29 Ibid., Annex, para. IV.5. The latter is a reference to a decision to the effect 'that the use of the mechanisms shall be supplemental to domestic action and that domestic action shall thus constitute a significant element of the effort made by each Party included in Annex I to meet its quantified emission limitation and reduction commitments': Kyoto Protocol (2005), *Decision 2/CMP.1, Principles, Nature and Scope of the Mechanisms Pursuant to Articles 6, 12 and 17 of the Kyoto Protocol*, FCCC/KP/CMP/2005/8/Add.1, para. 1.
30 Kyoto Protocol (2005), *Decision 27/CMP.1*, Annex, para. IV.6.

are open to the Facilitative Branch: (1) 'provision of advice and facilitation of assistance' to a party (i.e. any party) regarding the implementation of the Protocol; (2) 'facilitation of financial and technical assistance' to 'any Party concerned', including technology transfer and capacity-building 'from sources other than those established under the Convention and the Protocol for the developing countries'; (3) facilitation of financial and technical assistance, including technology transfer and capacity-building, 'taking into account' article 4.3/4.5 of the FCCC (relating to finance and technology transfer to non-Annex I parties); and (4) formulation of recommendations to 'the Party concerned' that 'take into account' article 4.7 of the FCCC (to the effect that the implementation of non-Annex I commitments under the Convention will depend on Annex I parties discharging their obligations on finance and technology transfer).[31]

What all this means is far from clear. Two of the four 'consequences' (the second and the fourth) relate to questions of implementation, because 'Party concerned' is defined in the rules as a party in respect of which a question of implementation has been raised.[32] The other two response types presumably can be used in any situation that has engaged the Facilitative Branch, whether it concerns a question of implementation or not. The second and third response types permit the Facilitative Branch to engage in facilitation of financial and technical assistance. However, the branch was not given a fund, or access to a fund or to another mechanism to achieve such an outcome. Therefore, these two response types exist only in theory.

Once a problem of implementation has found its way to one of the branches, certain procedures are common to, and must be followed by, both. The 'Party concerned' is entitled to designate persons to represent it for the duration of the process.[33] Each branch must base its deliberations on information obtainable from a limited number of sources. Sources include information provided by the party concerned, reports of Expert Review Teams, and reports of the Protocol party meetings and the subsidiary bodies of the Convention and the Protocol.[34] Relevant factual and technical information from 'competent' IGOs and NGOs may also be considered.[35] Each branch may seek expert advice in reaching a decision,[36] and the party concerned is to be given an opportunity to comment on any branch decision.[37]

Scholarly discussion of the Facilitative Branch has been rare, probably because the branch has remained largely inactive. The definition of a big part of its role in negative terms (i.e. that it is to address questions of implementation outside the mandate of the Enforcement Branch) has not been picked up as a fault,[38] and

---

31 Ibid., Annex, section XIV.
32 Ibid., Annex, para. VI.2.
33 Ibid., Annex, para. VIII.2.
34 Ibid., Annex, para. VIII.3.
35 Ibid., Annex, para. VIII.4.
36 Ibid., Annex, para. VIII.5.
37 Ibid., Annex, para. VIII.8.
38 E.g. Geir Ulfstein and Jacob Werksman (2005), 'Towards Hard Enforcement', p. 45; Jutta Brunnée (2012), 'Promoting Compliance with Multilateral Environmental

the absence of a trigger for the early-warning function has almost gone without comment. Lefeber, who was a member of the Enforcement Branch at the relevant time, writes that in the first two years of the Compliance Committee's operation (2006–2007) there was debate within the Facilitative Branch about whether its mandate allowed it to take action without the submission of a question of implementation.[39] However, the branch was not able to resolve the issue about the mechanism by which it should provide advice and facilitation. The internal debate was to continue, but it has still not been resolved.[40]

The performance of the Facilitative Branch is discussed in the second half of this chapter (Section 3.2.2), while in this section I focus the discussion on its rules. Suffice to say that, by 2011, the branch was facing an existential crisis and had still not clarified its mandate. In a meeting that year, branch members provided a new interpretation of the branch's rules, with the effect of expanding the branch's powers. Plausibility of textual construction appears not to have been the members' primary consideration in reaching this outcome. Their interpretation turns on reading the technical concept of 'question' in a provision on the Facilitative Branch's mandate as connoting *any kind* of 'issue':[41]

> the branch considered that the reference to having to 'take into account the circumstances pertaining to the questions before it' should *not* be interpreted to necessarily refer to questions of implementation. It is rather a reference to the *issues* before it, which could include questions of implementation.[42]

The branch members failed to mention that every other time the rules refer to a 'question' they clearly mean a question of implementation.[43]

Having established a new mandate over 'issues' in general, the Facilitative Branch proceeded to postulate a corresponding triggering mechanism. It was decided that an intervention by the branch could be triggered by 'issues' found in the reports of Expert Review Teams.[44] The reasoning may seem arbitrary and the result vague, but the branch now had a mechanism for its advisory/supportive

---

Agreements', in *Promoting Compliance in an Evolving Climate Regime*, ed. Jutta Brunnée, Meinhard Doelle, and Lavanya Rajamani, p. 51.

39 René Lefeber (2009), 'The Practice of the Compliance Committee under the Kyoto Protocol to the United Nations Framework Convention on Climate Change (2006–2007)', in *Non-Compliance Procedures and Mechanisms and the Effectiveness of International Environmental Agreements*, ed. Tullio Treves et al., p. 311.

40 See Compliance Committee (2012), *Annual Report*, FCCC/KP/CMP/2012/6, p. 14.

41 Last sentence of para. IV.4 in Kyoto Protocol (2005), *Decision 27/CMP.1*, Annex.

42 Compliance Committee (2011), *Annual Report*, FCCC/KP/CMP/2011/5, paras 48–49, emphasis added.

43 This is also true of the phrase *question before it*—as in 'The relevant branch shall undertake a preliminary examination of questions of implementation to ensure that, except in the case of a question raised by a Party with respect to itself, the *question before it*...': Kyoto Protocol (2005), *Decision 27/CMP.1*, Annex, para. VII.2, emphasis added.

44 Compliance Committee (2011), *Annual Report*, paras 50–51. This argument relied on section VII of the rules on procedures and mechanisms (Decision 27/CMP.1, Annex),

and early-warning functions. The branch conceded that it could not proceed with potential issues in 'the absence of procedures and the need to provide procedural safeguards to Parties'.[45] It resolved to continue to 'clarify its practice and/or procedures on how to discharge its responsibilities'.[46]

In the following year, 2012, the Facilitative Branch agreed on 'indicative working arrangements' for its provision of advice and facilitation. It called the arrangements a 'work in progress', to be tested and reviewed in practice 'as the branch considers its first cases'.[47] None of these qualifications change the nature of the branch's intended action, namely to create a way to call before it individual states on the basis of an allegation that they are on a path to non-compliance.

The plenary of the Compliance Committee, which includes the members of the Enforcement Branch, was in two minds about the Facilitative Branch's indicative working arrangements. By that point in time, the Enforcement Branch had completed several cases against states (see Section 3.2.3, below) on the basis of its own detailed rules of procedure, which had been drawn up by the Conference of the Parties to the Protocol. Its members presumably were not impressed by the creative, in-house process developed by the Facilitative Branch off its own bat.[48] The plenary meeting of the Committee declared that it was necessary for the Facilitative Branch to further enhance the transparency and due process of its indicative working arrangements, giving due consideration, among other things, to the need to 'systematically examine all reports of expert review teams, to ensure fair and equal treatment of all Parties', develop criteria for deciding whether to address an issue of early warning (as well as provide a definition of 'early warning'), and to further clarify its approach to the range of remedies it would utilize.[49]

As the first commitment period drew to a close, the modalities, if not the very purpose, of the Facilitative Branch were still unsettled. Nevertheless, the branch had

---

however, the relevant text does not refer to any questions other than questions of implementation.
45 Ibid., pp. 12–13.
46 Ibid., para. 53.
47 Compliance Committee Facilitative Branch (6–8 February 2012), *Report on the Eleventh Meeting*, CC/FB/11/2012/2, p. 2. The indicative working arrangements appear as an annex to this report. Early-warning intervention by the Facilitative Branch would be on the basis of 'sufficient information in [ERT reports] indicating potential non-compliance with commitments'. Concerning commitments under art. 3.1 of the Protocol (quantified mitigation limits), the branch would focus on emission projections and measures being taken by the party to address the potential non-compliance: ibid., p. 2. In the absence of formally established procedures for the early-warning function, the Facilitative Branch would apply, mutatis mutandis, procedures developed for dealing with questions of implementation: ibid., pp. 5–6.
48 The Enforcement Branch has its own procedures set out in section IX of the procedures and mechanisms (Decision 27/CMP.1, Annex), whereas the Facilitative Branch has only the general procedures of section VIII.
49 Compliance Committee Plenary (9 February 2012), *Report on the Tenth Meeting*, CC/10/2012/2, pp. 2–3. See also Compliance Committee (2012), *Annual Report*, p. 16.

received the Compliance Committee's tacit approval to proceed with developing its mandate. It was even suggested that inconsistencies in state reporting on the LULUCF sector was 'a possible starting point' for the advisory work of the Facilitative Branch.[50]

3.1.4 Powers of the Enforcement Branch

The functions of the Enforcement Branch are defined more carefully than those of the Facilitative Branch. It is responsible for determining whether an Annex I party has failed to comply with, first, the methodological and reporting requirements under articles 5.1/5.2 and 7.1/7.4 of the Protocol; second, the eligibility requirements under the Protocol's three flexibility mechanisms (articles 6, 12, and 17); and, third, the party's quantified emission-limitation commitment for a commitment period.[51] According to the ERT review rules, 'potential problems' under these headings, if not resolved at the ERT level, must be recast as questions of implementation and forwarded to the Enforcement Branch (via the Compliance Committee's Bureau).

Eligibility for participation in the Protocol's flexibility mechanisms requires compliance with the main methodological and reporting requirements under articles 5.1/5.2 and 7.1/7.4 of the Protocol, including those for national systems, national registries, and annual reporting of greenhouse gas emissions.[52] Therefore, the second category in the list above is but a special case of the first.

In connection with article 5.2 (first category above), the Enforcement Branch is tasked to decide whether an Annex I party's greenhouse gas emission inventory should be 'adjusted' in a case where an Expert Review Team has called for an upward adjustment of the estimated emissions (downward adjustments are not allowed) and the state concerned does not agree with the ERT. This too is called a question of implementation.[53]

---

50 Compliance Committee Enforcement Branch (7–8 and 10 February 2012), *Report on the Eighteenth Meeting*, CC/EB/18/2012/3, pp. 5–6.
51 Kyoto Protocol (2005), *Decision 27/CMP.1*, para. V.4.
52 Kyoto Protocol (2005), *Decision 2/CMP.1*, para. 5; and Kyoto Protocol (2005), *Decision 3/CMP.1, Modalities and Procedures for a Clean Development Mechanism as Defined in Article 12 of the Kyoto Protocol*, FCCC/KP/CMP/2005/8/Add.1, Annex, para. 31. These eligibility requirements for the CDM are repeated verbatim in the decisions setting up trading mechanisms under articles 6 and 17 of the Protocol. A party may also be excluded from the flexibility mechanisms if major errors in its reporting are uncovered, notwithstanding the fact that adjustments are effected to correct these errors: Kyoto Protocol (2005), *Decision 15/CMP.1, Guidelines for the Preparation of the Information Required under Article 7 of the Kyoto Protocol*, FCCC/KP/CMP/2005/8/Add.2, para. 3.
53 Kyoto Protocol (2005), *Decision 27/CMP.1*, Annex, para. V.5; Kyoto Protocol (2005), *Decision 22/CMP.1*, Annex, para. 80; and Compliance Committee (2012), *Annual Report*, pp. 13–14 ('With regard to the question of implementation relating to the disagreement about whether to apply adjustments...'). For the adjustment procedure, see Compliance Committee Enforcement Branch (3–4 July 2013), *Report on the Twenty-Third Meeting*, CC/EB/23/2013/3, Annex, pp. 5–6. The procedure was

Where the Enforcement Branch decides to proceed with a question of implementation, the party concerned may make written submissions to the branch and request a hearing to present its views. The branch has the power to call upon expert advice to supplement the information it has received from the ERT and any information it has received from the Annex I party. (The general procedure applying to both branches of the Compliance Committee was summarized in the previous section.)

In the case of a finding of non-compliance, the Enforcement Branch must apply 'consequences' (a euphemism for penalties) corresponding to the three kinds of non-compliance mentioned above.[54] (Adjustments to inventories are not regarded as consequences, since the party concerned is not required to take any action.)

In particular, where the non-compliance relates to methodological and reporting requirements, the branch must declare the state non-compliant and request it to submit a plan that will bring it into compliance.[55] The state's 'compliance action plan'[56] is subject to assessment (and acceptance or rejection) by the branch. The party must submit progress reports to the branch on the implementation of its plan on a regular basis.[57] This provision is often cited to illustrate the Enforcement Branch's 'automaticity'; however, close attention to the language reveals a discretionary element: 'it shall apply the ... consequences, taking into account the cause, type, degree and frequency of the non-compliance of that Party'.[58]

Where the non-compliance concerns the eligibility requirements for participation in the Protocol's flexibility mechanisms, the Enforcement Branch must suspend the party's eligibility to trade in the Protocol's emission allowances (AAUs, CERs, etc.).[59] The party may apply to the branch to have its eligibility to participate in the flexibility mechanisms restored.[60] The branch will normally await confirmation by an Expert Review Team that the non-compliance issue is resolved before reinstating the party's eligibility to trade.

The third 'consequence' available to the Enforcement Branch is relevant only once all reporting for a commitment period has been finalized. Where the branch has determined that the emissions of a party have exceeded its assigned amount for the commitment period, taking into account all emission allowances held by the party, the branch is to deduct from the party's assigned amount for the

---

approved in Kyoto Protocol (2013), *Decision 8/CMP.9, Compliance Committee*, FCCC/KP/CMP/2013/9/Add.1. The Enforcement Branch has a similar mandate where an ERT recommends a correction to the compilation and accounting database for the accounting of assigned amounts under art. 7.4 of the Protocol and the party concerned disagrees with the ERT: Kyoto Protocol (2005), *Decision 27/CMP.1*, Annex, para. V.5.

54 Kyoto Protocol (2005), *Decision 27/CMP.1*, Annex, para. V.6.
55 Ibid., Annex, para. XV.1.
56 For a description of which, see ibid., Annex, para. XV.2.
57 Ibid., Annex, para. XV.3.
58 Ibid., Annex, para. XV.1.
59 An exception applies to ERUs generated from a JI project hosted by the country, as well as to CERs forwarded by a developing country hosting a CDM project for the Annex I party in question.
60 Kyoto Protocol (2005), *Decision 27/CMP.1*, Annex, para. XV.4.

subsequent commitment period 1.3 times the emissions that were in excess in the earlier period. The party must also submit a compliance action plan and is to be temporarily blocked from selling any of its own AAUs to other parties.[61] At the time of writing (2014), finalization of the reporting for the first commitment period was not expected for another year. It seems safe to predict (see Table 4.5 in Chapter 4 and the discussion following it) that the non-compliance to which this third procedure corresponds will be avoided by all Annex I parties to the Kyoto Protocol. For these reasons there is little that needs to be said about this procedure, except to point out that it is potentially intrusive[62] and that state emissions during the first commitment period *might* have been different had the particular penalty not stood as a threat on the horizon throughout 2008–2012.[63]

3.2 Compliance system in practice

3.2.1 A system out of balance

In Chapter 2, I discussed the limited ability of Expert Review Teams to test the accuracy of greenhouse gas inventories.[64] In this section, I will examine a different problem involving ERTs, one which arises only in the context of the Kyoto Protocol's compliance system. The problem, in summary, is that whereas the facilitative role of ERTs under the FCCC's accountable reporting system is encouraged through regulation and practice without any resulting systemic difficulty, under the Protocol the same facilitative role undermines the functioning of both the Facilitative Branch and the Enforcement Branch.

When an ERT lists a reporting or institutional issue as a question of implementation in its annual review report on a state, action by the Compliance Committee is automatically triggered. It is a sign that the ERT and the state under review have failed to see eye-to-eye on a particular point. It elevates a behind-the-scenes factual dispute about a 'potential problem' into an open dispute that requires resolution by a higher authority. The question of implementation will be understood by all involved, as well as by the international community at large, as an allegation of non-compliance by the state with important regime rules. These are unpleasant consequences for a state.[65]

---

61 Ibid., Annex, para. XV.5.
62 See especially ibid., Annex, para. XV.6.
63 Thus the compliance system's theoretically 'most controversial aspect' (Jacob Werksman (2005), 'The Negotiation of a Kyoto Compliance System', in *Implementing the Climate Change Regime: International Compliance*, ed. Jon Hovi) will be remembered as its least controversial aspect in practice. As there will be no *third* commitment period in which to punish states with excess emissions in the second period, the system's one chance to generate controversy has passed.
64 See Section 2.3.2, above.
65 Greece and Canada were the first countries to have questions of implementation raised against them. They reacted with indignation. Canada, which at the relevant time had already declared that it did not intend to meet its emission-reduction target under the Protocol (Sebastian Oberthür and René Lefeber (2010), 'Compliance System Revisited', pp. 155–156), sought to have the record of the proceedings before the

Expert Review Teams are not as independent as some commentators have suggested.⁶⁶ Almost all reviewers have regular jobs in their national government's 'national system' (the state's greenhouse gas reporting authority), which gives them a role in the preparation of their country's national communications and greenhouse gas inventories. They thus serve on two rungs of the reporting-and-review ladder: for the most part, they are engaged in compiling their own country's reports, but once in a while they help to review those of other countries. They have an interest to see that their counterparts in other countries are playing fair, but they also have an interest not to come down too hard on them or expose them during the review of the national communication or inventory, for this could lead to similar treatment of their own country's reports.⁶⁷

These two matters (reputational costs for countries and a small circle of reviewers) are of little consequence for accountable reporting under the FCCC. A negative finding in an FCCC review report lies at the very end of the accountable reporting procedure under that treaty. It does not trigger any follow-up. The same negative finding in a Protocol review report could trigger a whole other, very public, procedure.

The upshot has been that ERTs have interpreted their mandate about whether to list a question of implementation for action by the Compliance Committee as discretionary. In this, they have been assisted by an imprecision in the applicable rules concerning which reporting elements are mandatory for states to follow and which are not.

A question of implementation has never arisen from a 'periodic review' of a national communication, and only eight such questions have resulted from an annual inventory review. It is clear from the way ERTs have been functioning that as long as they regard the Compliance Committee's involvement as unnecessary in the resolution of an infringement against the rules, whether mandatory or not,

---

Enforcement Branch reflect that there had been 'no need' to engage the non-compliance procedure, since very early on the process it had put right the identified problem. See Compliance Committee (2008), *Annual Report*, FCCC/KP/CMP/2008/5, para. 30 and Annex V. Greece's responses were also irate; see the documentation at the Kyoto Protocol's website, 'Question of Implementation – Greece', <http://unfccc.int/kyoto_protocol/compliance/enforcement_branch/items/5455.php>, especially the document listed as 'Further Written Submission of Greece' and dated 9 April 2008. See also Malgosia Fitzmaurice and Catherine Redgwell (2000), 'Environmental Non-Compliance Procedures and International Law', 31 *Netherlands Yearbook of International Law* 35, p. 64 ('It remains the case that the finding that a state has committed an internationally wrongful act as a matter of international law has significant psychological impact').

66 Geir Ulfstein and Jacob Werksman (2005), 'Towards Hard Enforcement', p. 43 ('The credibility of the ERTs lies first of all in their composition as a team of independent experts'). Oberthür and Lefeber have suggested that the Kyoto Protocol's compliance system has been designed to minimize political interference: Sebastian Oberthür and René Lefeber (2010), 'Compliance System Revisited', p. 140.

67 Taryn Fransen (2009), *Enhancing Today's MRV Framework to Meet Tomorrow's Needs: The Role of National Communications and Inventories*, p. 8.

an ERT will keep the issue under review at its own level without entering a formal finding of a question of implementation.

The integrity of ERTs is not in question here. Especially when conducting an annual review, which is about actual emissions and thus goes to the heart of the climate change regime, an ERT will question the states where there is evidence of significant underestimation. But other problems will not be aggressively pursued. The common practice (as with FCCC reviews) is to list the problems as recommendations for improvement for the next time around. If the improvements are not implemented in the course of the next review cycle, the ERT newly assigned to review the Annex I party (for its membership is required to be different from one review to the next), may relist the desired improvements for yet another cycle[68] or it may relinquish them as different issues catch its eye. Where a state can demonstrate that its emission accounts are improving from year to year, an ERT will generally be content to conceptualize reported gaps as improvement targets for later reporting years rather than as matters to be immediately corrected.

Because the ERTs, the state parties, and indeed the FCCC Secretariat, are anxious to avoid questions of implementation, functions that by design are assigned to the Compliance Committee are held back in an ad hoc manner at the level of Expert Review Teams.

The problem I have described has not gone unnoticed by the Compliance Committee. In 2011, the Committee's plenary 'recommended' (for it has no mandate to issue orders to Expert Review Teams) that ERT reports clearly state whether or not an identified problem relates to 'language of a mandatory nature', along with the reason for such a determination. The plenary also recommended that if an ERT decides not to list a question of implementation in relation to 'an unresolved problem pertaining to language of a mandatory nature', the ERT should give reasons for its decision.[69] The latter recommendation implicitly recognizes, of course, that ERTs have prevented questions of implementation from reaching the Compliance Committee. It also suggests that if ERTs are to continue this practice, they should at least provide a rationale for it. Another way of reading this incident is that the Compliance Committee is not condoning the practice but is trying to put an end to it without saying so explicitly. The plenary's report is ambiguous and could support either reading.

Like the Compliance Committee, the Facilitative Branch also attempted at one point to influence ERTs to be more explicit about their decisions. It had every reason to do so, considering that no questions of implementation have come its way since 2006, when a question of implementation assigned to the branch was left undecided (see next section). The Facilitative Branch recognized that ERTs

---

[68] Kyoto Protocol (2005), *Decision 22/CMP.1*, Annex, para. 48 ('All final review reports prepared by the expert review team, except for status reports, shall include the following elements: ... (b) ... (iii) An assessment of any efforts by the Party included in Annex I to address any potential problems identified by the expert review team during the current review or during previous reviews that have not been corrected').

[69] Compliance Committee (2011), *Annual Report*, pp. 7–8.

themselves have a strong facilitation function written into their mandate.[70] The branch members discussed holding a workshop with ERT lead reviewers to focus on 'the issue of consistency of reviews'. What the Facilitative Branch meant by this was that it wanted greater clarity from ERTs on how 'potential problems' were being managed at the ERT level, and in particular it wanted clarity on 'how mandatory language is used in ERT reports ... with respect to identifying questions of implementation and their resolution; and also how the reports might signal the risk of potential non-compliance and the need for early warning'.[71] The workshop went ahead, but it was held in closed session, so we are not privy to what was discussed.[72]

Compared with the Facilitative Branch, the Enforcement Branch has seen some action but it has not had a taxing workload. Its last question of implementation was received in May 2012.

The Enforcement Branch found an occasion to express its dissatisfaction with the review system, and it did so in relatively strong terms. In the course of its deliberations on the case of Bulgaria, in 2011, the branch

> noted with concern the lack of clarity in the 2010 [ERT review report on Bulgaria], which does not clearly explain why unresolved problems did not result in the listing of questions of implementation pursuant to paragraph 8 of the annex to decision 22/CMP.1 [on the guidelines for review under article 8 of the Protocol]. In particular, differing interpretations of this provision may lead to different conclusions as to whether an unresolved problem is required to be listed as a question of implementation. This reveals more systemic issues that concern the review process under Article 8 of the Kyoto Protocol and the compliance system as a whole, which require urgent attention.[73]

There is a suggestion here that the mandatory/non-mandatory problem could be to blame. However, the Enforcement Branch also leaves open the possibility that the ERT fudged its report on Bulgaria to avoid listing additional questions of implementation against the country.

3.2.2 Failure of the Facilitative Branch

The Facilitative Branch of the Compliance Committee could in theory provide 'early warning' of potential non-compliance with reporting or methodological

---

70 Compliance Committee Facilitative Branch (23 March 2013), *Report on the Thirteenth Meeting*, CC/FB/13/2013/2, p. 2 ('Given the expert review teams' role in assisting Parties in improving their reporting of information, the branch was interested in further exploring how it could exercise its [own] facilitative function').
71 Compliance Committee Facilitative Branch (22–23 October 2012), *Report on the Twelfth Meeting*, CC/FB/12/2012/3, pp. 4–5.
72 Compliance Committee Plenary (22–23 March 2013), *Report on the Twelfth Meeting*, CC/12/2013/3, p. 2.
73 Compliance Committee Enforcement Branch (4 February 2011), *Decision on Bulgaria Case*, CC-2010-1-17/Bulgaria/EB, para. 14.

obligations relating to greenhouse gas inventories, or indeed with emission-reduction obligations. In the early years of the branch's operation, it was widely expected that it would detect potential compliance problems and defuse them before they became actual problems.[74] However, the branch's powers have so far remained largely theoretical.

In May 2006, South Africa, on behalf of the Group of 77 and China, submitted a question of implementation[75] to the Compliance Committee in respect of 15 Annex I parties that had missed the submission deadlines for their national communications and progress reports on the implementation of various Kyoto Protocol commitments.[76] The Bureau referred the case to the Facilitative Branch. As Lefeber has summarized the outcome, 'the Facilitative Branch was neither able to take a decision to proceed nor a decision not to proceed with the question of implementation'.[77] It was, therefore, left unactioned. According to Lefeber, the impasse was due to flawed reasoning by certain members of the branch. He argues that the branch should have proceeded to decide the question.[78] It was the first and last question of implementation to go before the Facilitative Branch.

In late 2011, the Facilitative Branch considered that the in-depth review report on Italy's fifth national communication as well as the annual review report on Canada's 2008 inventory (submitted in 2010) 'point to potential problems in the fulfilment of these Parties' commitments', which the branch considered engaged its role on the promotion of compliance and provision of early warning of potential non-compliance.[79] After further deliberation, however, the branch concluded that the information available to it on Italy was not sufficient for it to engage in an early-warning exercise.[80] That still left the Canada case. Indeed, this case was calling out for attention since before the start of the first commitment period. For example, on 15 May 2006, Canada's environment minister publicly admitted at a meeting of the Protocol parties that Canada could not meet its emission-reduction target, for it was 'unachievable'.[81] It was a case of non-compliance foretold, and for this reason it should have been a test case for the Facilitative Branch from the

---

74 See, e.g., René Lefeber (2001), 'From The Hague to Bonn to Marrakesh and Beyond: A Negotiating History of the Compliance Regime under the Kyoto Protocol', 14 *Hague Yearbook of International Law* 25, p. 47; and Jan Klabbers (2007), 'Compliance Procedures', in *The Oxford Handbook of International Environmental Law*, ed. Daniel Bodansky, Jutta Brunnée, and Ellen Hey, p. 999.
75 This is an alternative triggering mechanism for questions of implementation; see Kyoto Protocol (2005), *Decision 27/CMP.1*, para. VI.1.b.
76 Compliance Committee (2006), *Annual Report*, FCCC/KP/CMP/2006/6, paras 19–22.
77 René Lefeber (2009), 'The Practice of the Compliance Committee', p. 314.
78 Ibid., pp. 314–315.
79 Compliance Committee Facilitative Branch (11-12 October 2011), *Report on the Tenth Meeting*, CC/FB/10/2011/3, p. 3.
80 Compliance Committee Facilitative Branch (6–8 February 2012), *11th Meeting*, pp. 2–3.
81 Jane Matthews Glenn and José Otero (2013), 'Canada and the Kyoto Protocol: An Aesop Fable', in *Climate Change and the Law*, ed. Erkki J. Hollo, Kati Kulovesi, and Michael Mehling, pp. 497–498.

very beginning. Instead, the Facilitative Branch took no action on Canada until the last year of the first commitment period.

In February 2012, the Facilitative Branch noted the concern expressed by the ERT reviewing Canada's fifth national communication, that the country would not be able to comply with its emission-reduction obligations under the Protocol. By this stage Canada had already submitted to the treaty's depositary a notification of withdrawal from the Protocol, which would make withdrawal effective on 15 December 2012. Despite this, the Facilitative Branch decided that the time had come to act. It reasoned that, 'for the time being, [Canada] remained a Party to the Protocol'.[82] In Section 3.1.3 above, I described how, by early 2012, the Facilitative Branch had agreed on 'indicative working arrangements' for its provision of advice and facilitation to Annex I parties. The branch decided to proceed with the Canada case on the basis of its draft procedures. As a first step, the chairperson of the branch would send a letter to Canada.[83]

The letter began with the observation that Canada's projected annual emissions for 2008–2012 were 21 per cent higher than 1990 levels, well above the country's emission limit of –6 per cent. The letter continued in a tone of contrived procedural propriety to note that there had been no indication in the ERT's review report on Canada about whether, or how, Canada planned to stay within its emission limit. The letter referred to the ERT's concern that Canada could become non-compliant. Canada's formal notification of withdrawal from the Protocol was acknowledged in the chairperson's letter, but was set aside with the officious remark that, legally, the party's obligations remained unaltered for the time being. The letter concluded with informing Canada that the Facilitative Branch had decided that it was seized of an early-warning issue with respect to the country, and that before proceeding further with the issue it wished to offer Canada the opportunity 'to engage in a dialogue with the branch' to clarify Canada's position.[84]

The Canadian government replied to the letter by underscoring at the outset that this was the first time that the Facilitative Branch had taken up an early-warning issue, and that its procedures for doing so were as yet incomplete. It went on to dismiss the branch's approach, with the argument that state compliance with emission-limitation obligations was not due to be assessed until long after Canada's withdrawal from the Protocol in December 2012 had become effective. 'On this basis, we are of the view that there is, therefore, little value in further engagement with the facilitative branch at this time.'[85] It was an embarrassing finale to an ill-considered initiative. By coming too late and being entirely without a point, the Facilitative Branch's decision to test its new procedure on Canada only managed to underscore the body's ineptness.[86]

---

82 Compliance Committee Facilitative Branch (6–8 February 2012), *11th Meeting*, p. 3.
83 Ibid., p. 3.
84 Compliance Committee Facilitative Branch (22–23 October 2012), *12th Meeting*, Annex.
85 Ibid., Annex.
86 Canada's imminent withdrawal from the Protocol was only half the story. In relation to emission targets, the Facilitative Branch conceivably could have played a useful

Earlier in this chapter I discussed the Facilitative Branch's lengthy meditation on its mandate and its subsequent attempt to interpret its allocated modalities and procedures so as to give it an effective role in international compliance. Certainly, the original design of the branch was poorly thought out and incomplete. This contributed to the branch becoming immediately moribund. However, the system's main flaw lies in its attempt to balance one facilitation body (the Facilitative Branch) on top of another (the Expert Review Teams) in the context of a political aversion to questions of implementation.

The experience with the Facilitative Branch of the Kyoto Protocol's compliance system suggests that a duplication of facilitation through the addition of a body at a level more remote from the underlying facts of each case is no improvement on the original system of accountable reporting under the FCCC.

3.2.3 Narrow influence of the Enforcement Branch

Of the Enforcement Branch's eight cases to date, the majority involved problems of institutional design that were soon repaired. Greece, Bulgaria, Romania, the Ukraine, and Lithuania were criticized for aspects of their national system; whereas in a case involving Canada, the national registry was at fault. The Croatia case arose from a dispute about that state's assigned amount, not from any misapplication of reporting or institutional rules.[87] Slovakia's case concerned a disagreement with an Expert Review Team about whether to apply an inventory adjustment, as well as a second question of implementation concerning incomplete or methodologically deficient accounting of emissions from two sectors. Both questions were said to spring from the same cause, namely the management of Slovakia's national system.[88] All eight cases have been resolved.[89]

As I noted earlier, questions of implementation have never arisen from national communications but only from inventory reports. This suggests that ERTs are placing a much greater emphasis on the accuracy of the reporting of greenhouse gases (and the functioning of national systems) than the less data-driven subject of national communications.

---

role in Canada's case early in the first commitment period. As time passed, however, Canada built up an emission profile that was increasingly difficult to change. By the time the branch began to reflect on its utility in late 2011, state policies, such as Canada's, had been fixed for the commitment period, and nothing could be done, realistically, in the final year of the period to significantly reduce the annual average of a country's emissions for 2008–2012. The Facilitative Branch's action on Canada was not only officious but illogical.

87 See Compliance Committee (2012), *Annual Report*, pp. 9–10.
88 Ibid., pp. 13–14.
89 The earlier Enforcement Branch cases are discussed in detail in Meinhard Doelle (2010), 'Early Experience with the Kyoto Compliance System: Possible Lessons for MEA Compliance System Design', 1 (2) *Climate Law* 237.

Once a case is before the Enforcement Branch, the branch begins to engage in facilitation, aimed at returning the country to compliance.[90] This is another way in which the facilitation function is shared within the Protocol's compliance system. The states that appear to have benefitted most from the experience are the five economy-in-transition cases (Bulgaria, Romania, Ukraine, Lithuania, and Slovakia), not counting Croatia (which was a case of legal interpretation of a rule). For them, the process worked in the manner of an external consultancy to fix systemic problems. The Enforcement Branch normally sought expert advice, both to find a solution and to confirm that it had been implemented. All experts were chosen from the FCCC Roster of Experts and had experience as ERT members.[91] Thus, even when a question was in the hands of the Enforcement Branch, most of the advice was coming from the ERT level. The branch provided little more than a forum to focus the attention of states on that advice.

States have responded with seriousness to the Enforcement Branch's proceedings against them. This is evident from the profiles of those assigned to represent the states at the branch's hearings. Lithuania sent a team comprising ten government ministers and senior public servants, plus a lawyer from a top legal firm.[92] At its second hearing it sent a team of six.[93] In Ukraine's case, eleven senior government officials plus an interpreter went to the first hearing,[94] and the same group was dispatched again to the second.[95] Slovakia was represented by seven officials; they included greenhouse gas experts and academics.[96] Romania sent nine high-level officials, including a lawyer and a greenhouse gas expert.[97] The response pattern is consistent with what we know about states being highly sensitive about being found in violation of agreed rules.

Where a system provides conditional benefits, such as optional emission trading under the Protocol, participants will be cut off from the benefits if they do not comply with the conditions of the special scheme. In the Protocol's case, the

---

90 For example, see the extent of facilitation invested by the Enforcement Branch in the Lithuania case (Compliance Committee Enforcement Branch (14–18 November 2011), *Report on the Sixteenth Meeting*, CC/EB/16/2011/2, pp. 4–6, and idem (9–14 July 2012), *Report on the Twentieth Meeting*, CC/EB/20/2012/2, pp. 7–8) as well as the Ukraine case (Compliance Committee Enforcement Branch (20–21 December 2011), *Report on the Seventeenth Meeting*, CC/EB/17/2011/2, pp. 3–4).
91 See, e.g., Compliance Committee Enforcement Branch (14–18 November 2011), *16th Meeting*, pp. 3–6; Compliance Committee Enforcement Branch (8–9 March 2012), *Report on the Nineteenth Meeting*, CC/EB/19/2012/2, p. 3; Compliance Committee Enforcement Branch (9–14 July 2012), *20th Meeting*, pp. 5–8; and Compliance Committee Enforcement Branch (3–4 July 2013), *23rd Meeting*, pp. 2–3.
92 Compliance Committee Enforcement Branch (14–18 November 2011), *16th Meeting*, pp. 3–4.
93 Compliance Committee Enforcement Branch (9–14 July 2012), *20th Meeting*, p. 6.
94 Compliance Committee Enforcement Branch (7–8 and 10 February 2012), *18th Meeting*, pp. 3–4.
95 Compliance Committee Enforcement Branch (8–9 March 2012), *19th Meeting*, pp. 2–3.
96 Compliance Committee Enforcement Branch (9–14 July 2012), *20th Meeting*, p. 3.
97 Ibid., p. 5.

Enforcement Branch has the authority, outlined earlier in this chapter, to suspend and reinstate eligibility to participate in the flexibility mechanisms. States that appear before the branch and are temporarily excluded from the flexibility mechanisms seem very keen to be reinstated as soon as possible. It might be thought that all that they really want is to regain the benefit of trade and potentially lessen the costs of compliance with the Protocol by buying or (for EIT countries, especially) selling emission allowances. The presence of multiple motives has always made explanation and prediction difficult in the social sciences. Considerable weight inevitably is given to how states account for their own conduct, and international law traditionally has relied heavily on this source. The evidence strongly suggests that states wish to comply with their accountable reporting obligations under the Kyoto Protocol because they reflect what international law requires and not solely for their own economic benefit.

In conclusion, it is difficult to assess the overall influence of the Enforcement Branch on state compliance with accountable reporting rules and the quality of state reports. It is possible that, in the absence of the branch, some states would have produced less complete or transparent accounts of their actions and emissions.[98] Such shortcomings would have been detected by Expert Review Teams under the regular FCCC review process and publicly highlighted as areas needing improvement. The difference is that they would not have been listed as questions of implementation.

3.3 Viability of the Protocol's compliance system

This chapter has surveyed the broad discretion available at the Expert Review Team level to report questions of implementation to the Compliance Committee. The existence of the discretion explains why the Protocol's compliance system has generated only eight cases for the Enforcement Branch and (in terms of cases originating in ERT reviews) none for the Facilitative Branch. The answer is that the ERTs have been pursuing facilitation efforts themselves.[99] There is a marked desire by all involved in the Protocol's process to avoid questions of implementation. At the negotiation stage, the major green NGOs favoured a hard-hitting compliance

---

[98] Lefeber assumes that the Compliance Committee's light workload is a good sign: René Lefeber (2009), 'The Practice of the Compliance Committee', p. 317 ('The Committee does not have much work and that must mean that the Kyoto Protocol, including the procedures and mechanisms relating to compliance, has made a good start'); see also George W. Downs (1998), 'Enforcement and the Evolution of Cooperation', 19 *Michigan Journal of International Law* 319, p. 331 ('The fact that punishments are rare can just as easily be interpreted as evidence that enforcement is operating effectively as that it is irrelevant').

[99] This has been noted by Meinhard Doelle (2010), 'Kyoto Compliance System', p. 259 ('The ERT process [is] not consistently bringing issues of implementation before the compliance committee. ... Whether the review by ERTs, in particular through in-country reviews, is sufficiently detailed and frequent for the credibility and integrity of the reporting system is unclear based on the experience to date').

system,[100] afflicted though this position was by the oxymoron that states would be willing to deliver and suffer heavy blows to themselves. The European Union did favour a strong compliance system, but not all Annex I parties agreed with it.[101] In its implementation, if not in its design, the Protocol compliance system has not been hard-hitting. ERT de facto facilitation has ensured that states only rarely will get into trouble with the Compliance Committee. By the second half of the Protocol's first commitment period, the Facilitative Branch was searching for ways to make a contribution to the compliance process. Its desire to elaborate its mandate led it to a tendentious reading of the applicable rules.

Over its eight years of operation, the Compliance Committee has averaged one case per year. No-one foresaw that the facilitative spirit and discretionary approach of Expert Review Teams to questions of implementation would undermine the Facilitative Branch and keep the Enforcement Branch barely active. The compliance system has evidently had no political appetite for being hard-hitting. The only alternative to this, of course, is facilitation. Yet facilitation can be handled well enough at the basic review (ERT) level, with the result that there is no significant role left for the specialized Protocol bodies at a higher level in the system.

Considering that the evidence on the workings of the Protocol's compliance system does not suggest that accountable reporting is definitely advantaged by the additional features introduced by that system, what have the state parties been saying about the model's long-term viability? In the ADP negotiations on the post-2020 agreement there is a consensus on the need for 'transparency for mutual trust, comparability and accountability' and for 'the need to take into account, and build on, the existing arrangements for measurement, reporting and verification [many of which] are just coming into effect and need to evolve'.[102] The latter is a reference to the IAR and ICA processes under the FCCC. Beyond this, positions diverge. At the COP in Warsaw in 2013, the LDCs and China argued for the features of the Protocol's compliance system to be transposed to the post-2020 agreement.[103] Presumably, they see it as continuing to apply only to Annex I parties. The United States, by contrast, said that existing reporting and review procedures under the FCCC are sufficient, as long as they are extended to

---

100 Steinar Andresen and Lars H. Gulbrandsen (2005), 'NGOs in Climate Compliance', p. 175.
101 See René Lefeber (2001), 'Negotiating History of the Compliance Regime', pp. 34–35. Lefeber's article shows that there was nothing inevitable about the Protocol's compliance system coming into being. The Sixth Conference of the Parties to the FCCC was not able to agree on it, and had to be suspended and resumed. In the resumed session, the parties again almost did not agree. With so little common ground on compliance, it would be implausible now to say that all states favoured a strong intepretation of the system they finally did agree on.
102 Ad Hoc Working Group on the Durban Platform for Enhanced Action (ADP) (13 August 2013), *Note by the Co-Chairs: Note on Progress*, ADP.2013.14.InformalNote, p. 2.
103 International Institute for Sustainable Development (2013), 'Warsaw Highlights: Friday, 15 November 2013', 12 (588) *Earth Negotiations Bulletin* 1, p. 2.

all parties according to capacity.[104] As I discussed at the end of Chapter 2, the US position appears to be informed by a deeper theory (or maybe assumption) that accountable reporting is itself an adequate compliance system.

'Managerialist' compliance theory could be read as suggesting that effective compliance management requires the establishment not only of transparent reporting but also a 'response system', such as the Protocol's compliance system: 'The information system must produce adequate and accurate information about actors' behaviours under the treaty. The managerial response system must then produce discriminating responses to different types of non-compliance.'[105] The production of 'discriminating responses' could, of course, take many forms, from recommendations in the review reports of Expert Review Teams, to question-and-answer sessions in the SBI forum under the IAR/ICA procedures.[106] It does not follow from managerialist principles that these responses are necessarily less effective than the approach represented by the Protocol's compliance system, which is itself, as I have explained, more facilitative than enforcement-oriented, and thus in practice no different from the other responses.[107]

There has been full cooperation of states with the Compliance Committee.[108] In this sense there has been compliance with the rules of the Kyoto Protocol's compliance system.[109] This further confirms the high level of state support for accountable reporting, which in turn supports the view that climate law is

---

104 International Institute for Sustainable Development (2013), 'Summary of the Bonn Climate Change Conference: 29 April – 3 May 2013', 12 (568) *Earth Negotiations Bulletin* 1, p. 13.
105 Abram Chayes, Antonia Handler Chayes, and Ronald B. Mitchell (1998), 'Managing Compliance: A Comparative Perspective', in *Engaging Countries: Strengthening Compliance with International Environmental Accords*, ed. Edith Brown Weiss and Harold K. Jacobson, p. 42.
106 See Michael Mehling (2012), 'Enforcing Compliance in an Evolving Climate Regime', in *Promoting Compliance in an Evolving Climate Regime*, ed. Jutta Brunnée, Meinhard Doelle, and Lavanya Rajamani, p. 197 (enforcement 'is today seen as also comprising all the actions undertaken by states [to] compel states ... this definition can include a range of flanking and alternative measures').
107 Bodansky makes this point more generally about MEAs: Daniel Bodansky (2010), *The Art and Craft of International Environmental Law*, p. 227 ('the procedures established by many recent agreements to identify and respond to cases of non-compliance have a primarily facilitative rather than an enforcement function [as] they seek to determine the cause of non-compliance and work with the state concerned to rectify the problem').
108 The only notable problem with the running of the Compliance Committee has been a frequent lack of quorum at meetings: Compliance Committee Enforcement Branch (22–23 March 2013), *Report on the Twenty-Second Meeting*, CC/EB/22/2013/3, p. 1; Compliance Committee Enforcement Branch (3–4 July 2013), *23rd Meeting*, p. 1; Compliance Committee Plenary (17–18 September 2013), *Report on the Thirteenth Meeting*, CC/13/2013/7, p. 2; and Compliance Committee (2013), *Annual Report*, FCCC/KP/CMP/2013/3, p. 10 ('the overall participation of members ... has declined over the last several years').
109 On whether states are legally obliged to comply with the decisions of the Compliance Committee, see René Lefeber (2001), 'Negotiating History of the Compliance Regime', pp. 50–53.

solidifying with respect to the report-and-review duty. The states that went before the Enforcement Branch did not evince any reluctance to report on emissions or have their reports reviewed; for the most part the problems were caused by reporting institutions that had not been set up in strict accordance with the rules.

Nevertheless, even if all states have complied with the compliance system's rules, it is still not clear that this system has added much value to the FCCC's version of accountable reporting.[110]

As for a legal duty to comply with core-outcome obligations (the substantive rule that states must reduce emissions), one looks in vain to the Protocol's compliance system for evidence of a normative development in this area. The Compliance Committee has not dealt with substantive obligations, and probably never will.[111] A more accurate term for the Committee, in retrospect, would have been the Uniform Reporting Committee.

Brunnée and Toope have noted 'the presence of strong shared understandings and the strong adherence to legality in the case of the regime's procedural aspects, and fragility of its substantive aspects'.[112] This difference can be explained from a legal perspective through the fact that the FCCC's rules on accountable reporting are, and always have been, directed at states individually; by contrast, the FCCC's general mitigation rule has always subsisted in its communal form, creating only implied, incidental obligations at the state level. This last point is the focus of the next chapter.

---

110 Thus the jury is still out on Oberthür's claim that 'Compliance mechanisms ... constitute a key factor strengthening the bindingness and effective implementation of international agreements' (Sebastian Oberthür (2014), 'Options for a Compliance Mechanism in a 2015 Climate Agreement', 4 (1–2) *Climate Law* 30, p. 31), at least as far as it applies to the climate change regime.

111 The IPCC's comment, that 'Despite the Kyoto Protocol's compliance system ... it is difficult in practice to enforce the Kyoto Protocol's targets because of the lack of a legal authority with enforcement powers, and the weakness of possible sanctions relative to the costs of compliance', therefore seems misconceived: IPCC (2014), *Climate Change 2014: Mitigation of Climate Change: Working Group III Contribution to the Fifth Assessment Report of the Intergovernmental Panel on Climate Change (Final Draft)*, ch. 13, p. 62. See also Jutta Brunnée and Stephen J. Toope (2010), *Legitimacy and Legality in International Law: An Interactional Account*, p. 195 ('the compliance mechanism of the Kyoto Protocol ... through its practice, can reinforce the protocol's substantive requirements').

112 Jutta Brunnée and Stephen J. Toope (2010), *Legitimacy and Legality*, p. 203.

4 State compliance with emission-limitation obligations

The last two chapters dealt with compliance matters that are facilitative to the central aim of the climate change regime, namely to have states make deep cuts in their greenhouse gas emissions. As I argued in those chapters, the obligation to engage in accountable reporting, while a step removed from the obligation to engage in emission reductions, is a case (a rare case) of hardened climate law, and is the only area of climate law where state compliance is high.

The climate change regime has available to it, in theory, a straightforward solution to the problem it sets out to confront. The solution is that states collectively are to stay within an overall emission budget; each state individually is to stay within *its* allocated budget; verification checks will periodically be made of state emission reports, and mismatches corrected; and when the global budget is updated by the IPCC, state budgets are to be adjusted accordingly under the guidance of an equitable formula. The regime is still a long way from realizing this ideal arrangement. Even proposals that meet the ideal half way seem improbable in the short term.[1]

The present chapter surveys the existing rules on emission limitation under the FCCC and the Protocol, and in the second part considers the extent of state compliance with these rules. It concludes with a discussion of the next steps necessary to develop the substantive law.

4.1 State obligations

4.1.1 Framework Convention

The FCCC was the first treaty to impose emission-reduction obligations on states. It contains two such obligations, one formulated generally and applying to all states, the other expressed concretely and applying to Annex I states only. The effect of the former is ongoing, whereas the latter expired in 2000. The short-lived rule was one of treaty law, whereas the continuing general rule is close to attaining the status of customary international law.[2] Binding though it is, it affects states communally

---

1 See Marco Grasso and J. Timmons Roberts (2014), 'A Compromise to Break the Climate Impasse', 4 *Nature Climate Change* 543.
2 Farber argues for 'a moral duty to impose reasonable curbs on its future emissions': Daniel A. Farber (2008), 'The Case for Climate Compensation: Justice for Climate

rather than individually,³ and therefore is less precise than the rule on accountable reporting discussed in Chapter 2, but it is climate law all the same.

The 'general' and 'specific' mitigation rules of the FCCC should be interpreted within the context of the whole treaty, with respect to each other, and taking into account subsequent decisions, pronouncements, and conduct of the FCCC parties. Because almost every country in the world is an FCCC party, the 'conduct of FCCC parties' is in essence that of the community of states.

The general mitigation rule is derived from the 'ultimate objective' of the FCCC, 'to achieve ... stabilization of greenhouse gas concentrations in the atmosphere at a level that would prevent dangerous anthropogenic interference with the climate system ... within a time frame sufficient to allow ecosystems to adapt naturally to climate change, to ensure that food production is not threatened and to enable economic development to proceed in a sustainable manner'.⁴ The rule prohibits state conduct that would amount to 'dangerous anthropogenic interference'. The test for this was clarified by FCCC parties to mean an increase in global average surface temperature of 2°C or more above pre-industrial levels.⁵

Change Victims in a Complex World', 2008 (2) *Utah Law Review* 377, p. 377. The communal legal duty in the FCCC is built upon this moral duty. No state has questioned what I call the general mitigation rule, although states have hardly embraced it. It represents a fundamental norm that is expressed in a great variety of ways. For a recent example, see US EPA (2014), *Carbon Pollution Emission Guidelines for Existing Stationary Sources: Electric Utility Generating Units: Proposed Rule*, p. 19 ('This proposal is an important step toward achieving the GHG emission reductions needed to address the serious threat of climate change. GHG pollution threatens the American public by leading to potentially rapid, damaging and long-lasting changes in our climate that can have a range of severe negative effects on human health and the environment'). I am conscious of the argument that customary law must be constituted by a formal act, such as a judicial decision, and that until such an act occurs there is no customary law, as such, but only a source of law. For the reasons given in the main text, the customary-law argument for climate law is premature.

3 Mehling calls laws in communal form 'commitments to the international community': Michael Mehling (2012), 'Enforcing Compliance in an Evolving Climate Regime', in *Promoting Compliance in an Evolving Climate Regime*, ed. Jutta Brunnée, Meinhard Doelle, and Lavanya Rajamani, p. 201 ('most commitments under the [FCCC] are not owed to a particular state but to the international community, or, in the case of commitments relating to technology transfer and financial assistance, to the entirety of developing countries'). Lefeber's apparently contrary view, that 'the obligation to mitigate climate change is not a collective obligation, but is binding upon individual developed countries' (René Lefeber (2012), 'Climate Change and State Responsibility', in *International Law in the Era of Climate Change*, ed. Rosemary Rayfuse and Shirley V. Scott, p. 330), is not accurate; the obligation has both collective and individual aspects, as I explain below.
4 FCCC, art. 2.
5 FCCC (2011), *Decision 2/CP.17, Outcome of the Work of the Ad Hoc Working Group on Long-Term Cooperative Action under the Convention*, FCCC/CP/2011/9/Add.1, part II.A, preamble. The FCCC bases its interpretation of 'dangerous' on IPCC advice. For that, see, most recently, IPPC (2014), *Climate Change 2014: Impacts, Adaptation, and Vulnerability: Working Group II Contribution to the Fifth Assessment Report of the Intergovernmental Panel on Climate Change: Summary for Policymakers*, pp. 12, 14, and 18.

88 *Emission-limitation obligations*

The general mitigaton rule implies 'deep cuts in global greenhouse gas emissions' and that all parties must take 'urgent action' to meet the 2°C goal.[6] Informally, the rule has also come to mean that 'stabilization' must be achieved by 2100.[7]

As of 2010, warming stood at 0.8°C above pre-industrial levels.[8]

The specific mitigation rule, or 'commitment' as it is called in the treaty, is that 'Each [Annex I party] shall adopt national policies and take corresponding measures on the mitigation of climate change, by limiting its anthropogenic emissions of greenhouse gases'.[9] The text proceeds to specify that each Annex I party is to limit its emissions in such a way as to 'return by the end of the present decade [i.e. by 2000] to earlier levels of anthropogenic emissions of carbon dioxide and other greenhouse gases'.[10] The 'limit' called for by the FCCC is, more precisely, a reduction, which is evident both from its use of the phrase 'return' to earlier levels and the fact that Annex I country emissions were increasing at the time the FCCC was signed, and were expected to continue to increase, year by year, if no action were taken.

The paragraph I have quoted from, which sets out the limitation/reduction obligation (article 4.2.a), does not clarify which 'earlier levels' are to be reached, but the next paragraph in the same article supplies the missing information: the 'aim' of Annex I parties is to return 'individually or jointly to their 1990 levels' of emissions.[11] Despite the option of 'jointly', each Annex I party individually must, according to the treaty, begin to reduce its emissions from earlier years, achieving that result by 2000 at the latest.

While the brunt of the specific mitigation commitment for the decade 1990–2000 is borne by Annex I parties, the FCCC also imposes mitigation commitments on members of the non-Annex I group. All parties, including developing countries, are to 'implement ... measures to mitigate climate change'.[12] This commitment, when considered in conjunction with the Convention's general mitigation rule (to stabilize greenhouse gas concentrations at a level that avoids dangerous climate change), strongly implies that non-Annex I parties must at least limit their emissions' rate of growth, and, somewhere along the way, also begin to reduce their emissions in an absolute sense. A purely logical construction of the article leads to the same conclusion, for the reduction commitment of Annex I parties would be pointless if non-Annex I parties were free under the treaty to increase their emissions without limit.

In Chapter 5, I will argue that the legal obligation imposed by the FCCC upon Annex I parties to financially assist non-Annex I parties to meet their mitigation

---

6 FCCC (2011), *Decision 2/CP.17*.
7 See Section 4.2.1, below.
8 IPCC (2013), *Climate Change 2013: The Physical Science Basis: Working Group I Contribution to the Fifth Assessment Report of the Intergovernmental Panel on Climate Change: Summary for Policymakers*, p. 3.
9 FCCC, art. 4.2.a.
10 FCCC, art. 4.2.a.
11 FCCC, art. 4.2.b.
12 FCCC, art. 4.1.b.

obligations, while having the appearance of procedural rule of climate law, is in fact an aspect of the general mitigation rule.

In summary, with respect to Annex I parties specifically, the FCCC imposes a mandatory emission-reduction obligation, but sets only one concrete target (1990 levels of emissions to be attained by 2000). The treaty allows for states to achieve that target individually or jointly. The FCCC is often pigeonholed as a framework treaty which is vague on targets,[13] yet the passages referred to above clearly imply that 1990 levels of emissions were never again to be exceeded by Annex I parties.[14] They also imply, in light of later elaboration, that states collectively are never to allow global average warming to reach 2°C (the general mitigation rule).

The general mitigation rule requires 'stabilization' of emissions in general, whereas the specific mitigation rule might be said to aim to stabilize Annex I emissions at 1990 levels. Both thus imply a reduction in emissions. However, this is where the commonalities end. The general mitigation rule embodies a test (the dangerousness or prevention test) that is not part of the specific mitigation rule—indeed, *could* not be part of it because the latter rule concerns only Annex I parties. The test gives the general mitigation rule a top-down character: emissions must be managed in such a way as to avoid dangerous climate change.[15] The rule mandates adherence to a global budget of permissible emissions. By contrast, the specific mitigation rule sets an arbitrary target. Almost certainly it represents nothing more than what Annex I parties considered they could accomplish by the year 2000. It is a bottom-up approach to mitigation. It is not concerned with prevention of dangerous warming.

It is sometimes said that states do not produce greenhouse gas emissions—industries and individuals do. The point opens up a reductionistic trap (viz. that the climate change problem is best managed, if at all, at the level at which it is caused). But the point is also a useful reminder of the fact that, if very drastic cuts in emissions were required immediately, most states could not deliver on them because their governments do not have that kind of power. Fortunately, the climate change problem is not yet so extreme that regular governmental authority is insufficient to stimulate the needed transformation. Given the global nature of climate change and the current depth of the problem, a solution by agreement at the state level is still the most appropriate response. This normative foundation strengthens the legal grip of the FCCC's rules: they not only address a world-

---

13 See e.g. Jacob Werksman (1999), 'Compliance and the Kyoto Protocol: Building a Backbone into a "Flexible" Regime', 9 *Yearbook of International Environmental Law* 48, p. 63 ('the convention's soft and ill-defined commitments'); Lavanya Rajamani (2012), 'Developing Countries and Compliance in the Climate Regime', in *Promoting Compliance in an Evolving Climate Regime*, ed. Jutta Brunnée, Meinhard Doelle, and Lavanya Rajamani, p. 378.
14 Thus Posner and Weisbach are wrong to say that 'the Framework Convention ... contains no binding limits on emissions': Eric A. Posner and David Weisbach (2010), *Climate Change Justice*, p. 61.
15 For a discussion of top-down versus bottom-up approaches in international climate policy, see William Hare et al. (2010), 'The Architecture of the Global Climate Regime: A Top-Down Perspective', 10 (6) *Climate Policy* 600, p. 601.

degrading, life-threatening problem, they also represent the only practical, reliable, and logical way to begin to tackle it.

The above is a summary of the FCCC's substantive climate law. The FCCC did not establish a formal compliance system to ensure that states comply with its rules, whether substantive (emission reductions) or procedural (accountable reporting). Even the specific mitigation rule, namely that Annex I emissions should be brought down to 1990 levels by 2000, was not supervised in any systematic way, although it was monitored by the Secretariat and discussed in the SBI forum. I will come to this later in the chapter when I consider whether the 2000 target was met.

The fact that the FCCC established no compliance system does not, of course, imply that compliance was not expected or was not to be taken seriously by the state parties. States must comply with regime rules whether or not there is a mechanism in place to respond to non-performance. Yet, what about the case where a legal obligation rests on states collectively, as is true of the Convention's general mitigation rule?

The most natural interpretation of the FCCC's communal rule is that it entails an additional, non-communal, individualistic aspect, namely that it places an obligation on states—that is, on each state individually—to participate in good-faith negotiations towards a burden-sharing agreement that would satisfy the collective obligation to bring emissions down to a level to avoid the 2°C threshold.[16] No other interpretation makes the rule effective. The general mitigation rule thus further entails that the original agreement must be developed into a new agreement on burden-sharing. Of course, it has long been clear that the FCCC was only ever the first step in the development of international climate law, but it has not often been remarked upon that what makes it the first step is that the law is communal in form. Yet another implication of the general mitigation rule is that states, while negotiating an individualization agreement, must conduct themselves consistently with the ultimate objective of the FCCC regime. The contrary again makes no sense: to postulate that individual states retain a right to accelerate their emissions without limit would be to render the FCCC ineffective and undermine the individualization project. This is because a state that is reducing its emissions year by year (whether absolutely or compared with business as usual) should find it easier to sign up to an agreement involving a global emission budget, than a developmentally equivalent state that has not made such a turn.

Individualized state obligation prior to the point when a burden-sharing agreement is signed is thus textually indistinct and attenuated, yet it exists by

---

16 The existence of an obligation on states to continue to participate in the climate change negotiations in good-faith to achieve the FCCC's objective and an individualization of the mitigation burden has been remarked upon by the International Law Association, which has proposed that a rule exists to that effect: Shinya Murase and Lavanya Rajamani (et al.) (2014), *Legal Principles Relating to Climate Change*, p. 18 ('Draft Article 5.2 … All States have a common responsibility to cooperate in developing an equitable and effective climate change regime applicable to all, and to work towards the multilaterally agreed global goal').

necessary implication of the original agreement. If the implied follow-up agreement were to come to life, its terms would regulate state action through state-specific emission targets. On this analysis, the early (pre-individualization) period requires both a good-faith participation in the negotiation of a burden-sharing agreement and a reasonable effort in the interim to reduce emissions and/or financially assist developing countries to do so. A state that does not observe these two implied rules would not be in compliance with the FCCC treaty. Hence the general mitigation rule, despite its communal form, could render individual state conduct non-compliant with the FCCC treaty in certain respects.

The FCCC obligation on Annex I parties to reduce their emissions to 1990 levels by 2000 was the first chapter in international efforts to get state emissions under control. It was followed by the 2008–2012 commitment period under the Kyoto Protocol, and this in turn was followed by emission limitations promised by a larger group of countries covering the period 2013 to 2020. State emissions in this last period are in fact controlled by two parallel regimes, making it legally more complex than anything that has gone before. (For a schematic representation, see Table 1.1 in Chapter 1.) The parallel regimes consist of, first, mitigation pledges by Annex I and some non-Annex I parties under the FCCC, a model introduced by the Copenhagen Accord in 2009 and absorbed into the FCCC rules at the 2010 Conference of the Parties in Cancun;[17] and, second, binding Annex I emission limits under the second commitment period of the Kyoto Protocol, agreed to at the Durban conference in 2011.[18] One and the same country can have different emission-reduction targets under these parallel regimes because of the differently constituted elements covered by the two regimes.

4.1.2 Kyoto Protocol period I

By comparison with the FCCC, the Protocol is more precise about the emission-reduction obligations of states. Like the FCCC's specific mitigation rule, it targets Annex I states only; it affirms the principle that, in general, the absolute ceiling of Annex I emissions should be the year 1990; and it allows states to meet their targets severally or jointly. The Protocol lowered the emission ceiling for Annex I parties compared with the FCCC's specific mitigation rule to a collective 5 per cent below 1990 levels for the 2008–2012 period. The question of how that period's 'budget' should be shared was settled in burden-sharing discussions. This last element is an innovation unique to the Kyoto Protocol.

---

17 See FCCC (2010), *Decision 1/CP.16, The Cancun Agreements: Outcome of the Work of the Ad Hoc Working Group on Long-Term Cooperative Action under the Convention*, FCCC/CP/2010/7/Add.1; and Section 2.2.5 in Chapter 2.
18 Kyoto Protocol (2011), *Decision 1/CMP.7, Outcome of the Work of the Ad Hoc Working Group on Further Commitments for Annex I Parties under the Kyoto Protocol at Its Sixteenth Session*, FCCC/KP/CMP/2011/10/Add.1.

In its original form (i.e. prior to the 2012 amendment that established a second commitment period)[19] the Protocol set quantified emission limitations for Annex I parties over a five-year commitment period:

> The Parties included in Annex I shall, individually or jointly, ensure that their aggregate anthropogenic carbon dioxide equivalent emissions of the greenhouse gases listed in Annex A do not exceed their assigned amounts, calculated pursuant to their quantified emission limitation and reduction commitments inscribed in Annex B and in accordance with the provisions of this Article, with a view to reducing their overall emissions of such gases by at least 5 per cent below 1990 levels in the commitment period 2008 to 2012.[20]

Because the specific mitigation rule expired in 2000, state mitigation action over the period 2001–2007 was not subject to any internationally pledged commitments. Legally, only the FCCC's general mitigation rule applied for this intervening period.

The Kyoto Protocol developed the FCCC's specific mitigation rule through the novel element of state-level quantification. By contrast, the FCCC's general mitigation rule, which is tied to the dangerousness test, was not further developed by the Protocol. Indeed, it could not have been, because a global budget of emissions that achieves non-dangerous stabilization could not be 'managed' through reductions in Annex I emissions alone. There have been suggestions that the Protocol stands for a top-down approach.[21] In fact, it is a continuation of the bottom-up tradition of the FCCC's specific mitigation rule, in which a few countries commit to a mitigation effort whose impact is essentially arbitrary and not in furtherance of the general, communal, rule.[22]

Supplementary emission-trading provisions (the flexibility mechanisms) in the Protocol meant that Annex I parties *could*, in fact, exceed their assigned amounts, as long as any excess was offset by emission allowances purchased from other Annex I parties or from CDM or JI projects. A country whose emissions exceeded its assigned amount for the commitment period would be allowed 100 days after the Expert Review Team's review of its final emission inventory to make up the

---

19 Kyoto Protocol (2012), *Decision 1/CMP.8, Amendment to the Kyoto Protocol Pursuant to Its Article 3, Paragraph 9 (Doha Amendment)*, FCCC/KP/CMP/2012/13/Add.1.
20 FCCC, art. 3(1).
21 E.g. Jutta Brunnée (2012), 'Climate Change and Compliance and Enforcement Processes', in *International Law in the Era of Climate Change*, ed. Rosemary Rayfuse and Shirley V. Scott, p. 311.
22 Jutta Brunnée and Stephen J. Toope (2010), *Legitimacy and Legality in International Law: An Interactional Account*, p. 167 ('The particular commitments taken on ... were clearly the result of a bargain rather than reflective of a principled agreement on the criteria for differentiation of individual states' emission targets').

shortfall.²³ The possibility of trade also meant, at least in theory, that no state that ended a commitment period with territorial emissions substantially below the prescribed level would be left uncompensated for its achievement. The excess emission allowances could be sold (assuming, of course, that the market at the time was not flooded with other states' unwanted allowances).

The Protocol's system to ensure state compliance with mitigation targets is a second major difference with the FCCC. As discussed in Chapter 3, if there is non-compliance with an Annex I party's emission limit after all of the state's allowance holdings are accounted for, the Enforcement Branch is to deduct 1.3 times the excess tonnes, as a penalty, from the state's assigned amount for the subsequent commitment period. The non-compliant state is also to submit a compliance action plan and be suspended, for an indefinite period, from emission trading under article 17 of the Protocol.²⁴

4.1.3 'Nationally appropriate mitigation commitments and actions'

I now proceed to the mitigation rules for the 2013–2020 period, beginning with the scheme under the FCCC. The following terminology applies: Annex I parties are required to implement 'quantified economy-wide emission-reduction targets' during this period.²⁵ Another term that has been used is 'nationally appropriate mitigation commitments or actions'.²⁶ It is an umbrella term for mitigation efforts for both Annex I and non-Annex I parties for the period 2013–2020. However, the first term is the one usually preferred for Annex I parties, because it mentions 'targets'. Developing-country parties, having resisted as much as possible being associated with targets, or indeed 'commitments', have instead accepted nationally appropriate mitigation actions (NAMAs).²⁷

In the wake of the negotiations under the Bali Action Plan, and adopting the language and timeframe of the Copenhagen Accord ('targets for 2020'),²⁸ Annex I parties agreed in 2010 to take on economy-wide emission-reduction targets for the new period.²⁹ The targets are self-determined, and many are qualified by conditions. A work programme was established under the SBSTA to 'clarify' the targets, and in particular to assess the 'comparability' of state effort that the various

---

23 Kyoto Protocol (2005), *Decision 27/CMP.1, Procedures and Mechanisms Relating to Compliance under the Kyoto Protocol*, FCCC/KP/CMP/2005/8/Add.3, section XIII.
24 Ibid., section XV.5.
25 FCCC (2010), *Decision 1/CP.16*, para. 36. These targets were originally listed in FCCC Secretariat (7 June 2011), *Compilation of Economy-Wide Emission Reduction Targets to Be Implemented by Parties Included in Annex I to the Convention*, FCCC/SB/2011/INF.1/Rev.1.
26 FCCC (2010), *Decision 1/CP.16*.
27 Ibid., para. 49. These are contained in FCCC Secretariat (18 March 2011), *Compilation of Information on Nationally Appropriate Mitigation Actions to Be Implemented by Parties Not Included in Annex I to the Convention*, FCCC/AWGLCA/2011/INF.1.
28 FCCC (2009), *Decision 2/CP.15, Copenhagen Accord*, FCCC/CP/2009/11/Add.1, para. 4.
29 FCCC (2010), *Decision 1/CP.16*.

Table 4.1. Quantified economy-wide emission-reduction targets, or pledges, by Annex I parties to the FCCC for the period 2013–2020. The asterisk (*) indicates states that are not participating in the Kyoto Protocol's second commitment period. A state's Protocol (KP) target for the second commitment period is shown in parentheses only where it differs from its FCCC pledge.[a]

EU Annex I party	Pledged action
European Union	*EU target*: 20% reduction in emissions from 1990 levels by 2020. An effort-sharing agreement determines the targets of individual member states.[b] A reduction of 30% is conditional on strong global action.
The EU target applies to: Austria, Belgium, Bulgaria, Croatia, Cyprus, the Czech Republic, Denmark, Estonia, Finland, France, Germany, Greece, Hungary, Ireland, Italy, Latvia, Lithuania, Luxembourg, Malta, the Netherlands, Poland, Portugal, Romania, Slovakia, Slovenia, Spain, Sweden, and the United Kingdom.	
The following four EU countries have a domestic 2020 target of greater ambition than the EU target:	
Denmark, Germany, and Sweden	Domestic target to reduce emissions by 40%.
United Kingdom	Statutory domestic target requires reduction of at least 34%.

Non-EU Annex I party	Pledged action, with conditions shown, where applicable
Australia	5% reduction in emissions from 2000 levels by 2020. Reduction of 15% or 25% is conditional on stronger levels of global action. (*KP target:* 2% below 2000 levels, which equals 0.5% below 1990 levels.)
Belarus	Conditional target of 5–10% reduction in emissions from 1990 levels by 2020. (*KP target:* 12% below 1990 levels.)
Canada*	17% reduction in emissions from 2005 levels by 2020.
Iceland	15% reduction in emissions from 1990 levels by 2020. Reduction of 30% is conditional on the EU adopting the 30% target.
Japan*	3.8% reduction in emissions from 2005 levels by 2020. Represents 3.1% increase in emissions from 1990 levels.
Kazakhstan	15% reduction in emissions from 1992 levels by 2020. (*KP target:* 5% below 1990 levels.)
Liechtenstein	20% reduction in emissions from 1990 levels by 2020. Reduction of 30% is conditional on strong global action. (*KP target:* 16% below 1990 levels.)
Monaco	30% reduction in emissions from 1990 levels by 2020. (*KP target:* 22% below 1990 levels.)
New Zealand*	5% reduction in emissions from 1990 levels by 2020. Reduction of 10% or 20% is conditional on stronger levels of global action.

Norway	30% reduction in emissions from 1990 levels by 2020. Reduction of 40% is conditional on strong global action. (*KP target*: 16% below 1990 levels.)
Russia*	15% reduction in emissions from 1990 levels by 2020. Reduction of up to 25% is conditional on certain forestry accounting rules being accepted and action by major emitters.
Switzerland	20% reduction in emissions from 1990 levels by 2020. Reduction of 30% is conditional on strong global action. (*KP target:* 15.8% below 1990 levels.)
Turkey	No pledged action.
Ukraine	Conditional target of 20% reduction in emissions from 1990 levels by 2020. (*KP target:* 24% below 1990 levels.)
United States*	Provisional target of 17% reduction in emissions from 2005 levels by 2020.

Notes:
a See Kyoto Protocol (2012), Decision 1/CMP.8.
b *Decision No. 406/2009/EC of the European Parliament and of the Council of 23 April 2009 on the Effort of Member States to Reduce their Greenhouse Gas Emissions to Meet the Community's Greenhouse Gas Emission Reduction Commitments up to 2020.* Individualized state limits are shown in Annex II to the decision.

Source: Adapted from FCCC Secretariat (2013), *Quantified Economy-Wide Emission Reduction Targets by Developed Country Parties to the Convention: Assumptions, Conditions, Commonalities and Differences in Approaches and Comparison of the Level of Emission Reduction Efforts*, FCCC/TP/2013/7, paras 17f.

proposals represent, taking into account the differences in the parties' national circumstances. The SBSTA was tasked to report on the outcome of this work programme in November 2014.[30] Implementation of the targets is nevertheless to proceed 'without delay'.[31]

The Annex I targets for 2013–2020 are shown in Table 4.1. To reiterate, pledges are 'voluntary' and are to be distinguished from the legally binding targets of those parties participating in the second commitment period of the Kyoto Protocol (which covers the same years). The non-binding FCCC targets are in some cases less conservative than the corresponding Protocol targets.[32] Where a state has put forth more than one target, the less ambitious of these targets (as shown in the table) is in general the unconditional one, whereas higher targets are mostly made subject to conditions. A small number of parties actually qualified their sole target

---

30 FCCC (2012), *Decision 1/CP.18, Agreed Outcome Pursuant to the Bali Action Plan*, FCCC/CP/2012/8/Add.1, paras II.A.8 and 9.
31 FCCC (2013), *Decision 1/CP.19, Further Advancing the Durban Platform*, FCCC/CP/2013/10/Add.1, para. 4(b).
32 The difference in some cases relates to whether the FCCC target is for the year 2020 alone or is, like the Protocol's target, for the whole of the 2013–2020 period. For example, in Australia's case it is the former whereas in the EU's case it is the latter. In other cases, e.g. Iceland and Monaco, the position is ambiguous. Note that an EIT country, such as Belarus or the Ukraine, whose emissions in the wake of economic transition are expected to catch up with their 1990 emission levels, tends to commit to deeper *period* reductions (Protocol) and shallower *2020* reductions (FCCC).

96 *Emission-limitation obligations*

(or range of targets) so that their pledges are fully conditional, or, as the United States describes its own target, 'provisional'.[33]

That the FCCC targets for Annex I parties for 2020 are not legally binding follows from the fact that the Convention does not provide a mechanism for them to be legally binding short of creating a new treaty (protocol) for that purpose, which the FCCC parties have not done. To say that the targets are 'voluntary' has approximately the same meaning as 'not legally binding' and possibly also hints that they were not imposed on states from above, and perhaps also that they are ad hoc. However, there is no doubt that all Annex I parties were under a legal obligation to make mitigation pledges. In the event, all of them except for Turkey[34] did so. As discussed below, the source of this obligation can be traced to a combination of the FCCC's general and specific mitigation rules.

I now proceed to consider the developing-country NAMAs. The Bali Action Plan proposed the adoption of NAMAs for developing countries 'supported and enabled by technology, financing and capacity-building, in a measurable, reportable and verifiable manner'.[35] By 2010, this had evolved into a requirement that developing countries '[aim] at achieving a deviation in emissions relative to "business as usual" emissions in 2020'.[36] This requirement remained conditioned on the provision of 'support' from Annex I parties. By mid-2013 (six months into the new period), 57 non-Annex I parties (about a third of developing countries) had submitted NAMAs.[37] They are not all of the same kind. A few are short and to the point—similar to the Annex I pledges. Others are detailed down to the project level and have the appearance of applications for funding. In an attempt to clarify them, the FCCC parties established a work programme under the SBI. A report on the programme's outcome was due in November 2014.[38]

As I noted in the introductory chapter to this book, mitigation commitments can range in type from weak to strong. I distinguished three main types: a reduction in the emission intensity of a state's economy, leading to a reduction in emissions compared with a hypothetical future trajectory in which emission intensity is not interfered with; a trajectory lower than the business-as-usual trajectory without specification of a quantified change in any indicator that is known to be a key

---

33 For an analysis of conditions and assumptions behind Annex I pledges, see FCCC Secretariat (2013), *Quantified Economy-Wide Emission Reduction Targets by Developed Country Parties to the Convention: Assumptions, Conditions, Commonalities and Differences in Approaches and Comparison of the Level of Emission Reduction Efforts*, FCCC/TP/2013/7.
34 Turkey is in a situation 'different from that of other Parties included in Annex I to the Convention' (FCCC (2001), *Decision 26/CP.7, Amendment to the List in Annex II to the Convention*, FCCC/CP/2001/13/Add.4, para. 3), and is not required to adopt mitigation targets under the FCCC or the Kyoto Protocol.
35 FCCC (2007), *Decision 1/CP.13, Bali Action Plan*, FCCC/CP/2007/6/Add.1, para. 1.b.ii.
36 FCCC (2010), *Decision 1/CP.16*, para. 48.
37 Subsidiary Body for Implementation (2013), *Compilation of Information on Nationally Appropriate Mitigation Actions to Be Implemented by Developing Country Parties*, FCCC/SBI/2013/INF.12/Rev.2.
38 FCCC (2012), *Decision 1/CP.18*, para. II.B.19.

Table 4.2 Nationally appropriate mitigation actions by non-Annex I parties to the FCCC for the period 2013–2020. Only countries with specific mitigation targets, as opposed to broadly indicated actions, are shown. Most developing countries presented their NAMAs as lists of projects for which financial support was sought, without an indication of the impact that such projects would have on a country's emissions. The pledges of the 16 non-Annex I parties that did present state-level targets have been divided here into three categories.

Non-Annex I party	Pledged actions are conditional upon receipt of international financial support (Israel is the only exception to this)
	I. Pledge expressed as absolute reduction from past level:
Antigua and Barbuda	25% reduction in emissions from 1990 levels by 2020.
Maldives	Carbon neutrality by 2020.
Marshall Islands	40% reduction in emissions from 2009 levels by 2020.
Moldova	25% reduction in emissions from 1990 levels by 2020.
	II. Pledge expressed as reduction in intensity from past level:
China	Reduction in emissions per unit of GDP of 40–45% from 2005 levels by 2020, and increase forest cover by 40 million hectares from 2005 levels by 2020.
India	Reduction in emissions per unit of GDP, excluding agriculture, of 20–25% from 2005 levels by 2020.
Malaysia	Reduction in emissions per unit of GDP of 40% from 2005 levels by 2020.
	III. Pledge expressed as reduction from hypothetical future pathway:
Brazil	36.1% to 38.9% reduction from projected business-as-usual emissions in 2020.
Chile	20% reduction from projected business-as-usual emissions in 2020 (on basis of projection made in 2007).
Indonesia	26% reduction from projected business-as-usual emissions in 2020.
Israel	20% reduction from projected business-as-usual emissions in 2020. (Financial support not sought.)
Korea (South)	30% reduction from projected business-as-usual emissions in 2020.
Kyrgyzstan	20% reduction from projected business-as-usual emissions in 2020.
Mexico	30% reduction from projected business-as-usual emissions in 2020.
Singapore	16% reduction from projected business-as-usual emissions in 2020.
South Africa	34% reduction from projected business-as-usual emissions in 2020.

Source: Adapted from Subsidiary Body for Implementation (2013), *Nationally Appropriate Mitigation Actions for Non-Annex I Parties.*

factor in the country's emissions; and an absolute reduction in emissions compared with a past baseline. All Annex I pledges are of the last type, which is the easiest to verify.[39]

Of the 57 developing countries with pledges for 2020, three chose an emission-intensity reduction, and nine a reduction from business-as-usual emissions without committing to a key indicator (Table 4.2). Only four states pledged absolute reductions. The remaining 41 states simply listed mitigation projects without claiming an expected reduction from business as usual.[40]

The four non-Annex I parties with absolute emission-reduction pledges have extremely low emissions already, so their pledges may be regarded as merely symbolic. The high-emitting developing countries (China, Brazil, etc.) all have pledges of the more speculative kind.

In 2012, the Conference of the Parties reiterated its 'invitation' to non-Annex I parties wishing to 'voluntarily' implement NAMAs to submit information to that effect.[41] In 2013 it 'urg[ed] each developing country Party that has communicated its nationally appropriate mitigation action to implement it and, where appropriate, consider further action'.[42] The language could be said to have shifted over time from hesitant to pressing.

The FCCC's mitigation regime for the 2013–2020 period contains two important new elements. First, mitigation actions are not limited to Annex I parties. All FCCC parties are encouraged to submit pledges. Many non-Annex I parties have done so. Even though most of their pledges are vague, some of the largest non-Annex I emitters have made mitigation pledges—and this is unprecedented. Second, the FCCC began a work programme to determine the comparability of the effort represented by the pledges. The programme was limited to the pledges of Annex I parties. Nevertheless, it is the first time that the FCCC parties as a whole have given serious attention to the issue of effort-sharing equity. (The Kyoto Protocol parties, but also the EU member states at the regional level, already have some experience with allocating burdens.)

The two new elements (broader participation and equitable sharing) are implicit in the FCCC's general mitigation rule. Because the obligation to reduce emissions

---

39 The development summarized here by Wang illustrates why absolute reduction commitments are necessary: 'With domestic coal demand in the United States expected to fall by 30% owing to the EPA rule [on emissions from coal-powered electricity plants], US coal firms—sitting on the largest recoverable reserves in the world—are pushing to increase exports to Asia, especially to China. Three new coal-export ports are being proposed for the Pacific coast, and are projected to ship up to 100 million tonnes of coal per year. The huge added supply to Asia will lead to cheaper coal and increased consumption.' (Qiang Wang, 2014, 'China Should Aim For a Total Cap on Emissions', 512 *Nature* 115.)

40 See Taryn Fransen (2009), *Enhancing Today's MRV Framework to Meet Tomorrow's Needs: The Role of National Communications and Inventories*, p. 14 ('Developing country NAMAs ... include diverse policies and measures, and are not encompassed by economy-wide emission targets ... This complicates attempts to define an appropriate [reporting and verification] framework for them').

41 FCCC (2012), *Decision 1/CP.18*, para. II.B.16.

42 FCCC (2013), *Decision 1/CP.19*, para. 4(f).

attaches to the community of states, each state must contribute in good faith to the development of an effort-sharing agreement through which specific emission reductions are allocated to individual states. There are differences, of course: the 2020 pledges were made *prior* to the launch of the effort-sharing negotiations, and the pledges are not referenced to a global emission budget but are instead ad hoc (they are as ad hoc as those of the FCCC's specific mitigation rule and those of the Protocol's first and second commitment periods). From a legal point of view, however, the 'law' for the period 2013–2020 is no longer simply a reiteration of the specific mitigation rule; it represents an elaboration, however slight, of the FCCC's general mitigation rule.

No compliance system within the FCCC oversees the Annex I targets or the non-Annex I NAMAs for the period to 2020. However, the accountable reporting rule clearly applies to this period and has been supplemented by the processes of IAR and ICA, as discussed in Chapter 2.[43]

4.1.4 Kyoto Protocol period II

Thirty-seven Annex I parties and the EU agreed to take on emission-reduction commitments for the Kyoto Protocol's second commitment period. Japan, Russia, and New Zealand did not join. Canada withdrew from the treaty before the first commitment period ended. The targets for the second period were to be decided in the AWG-KP negotiation process with reference to an overall (Annex I) reduction percentage from 1990 levels, but the parties failed to agree on a formula.[44] Targets were finally put forth by states in an ad hoc manner.

The transition from the first to the second commitment period was far from seamless. With five Annex I parties now on the sidelines (counting the United States), and the treaty doomed, the question of 'compliance' of the remaining parties with the Protocol's rules no longer seems so interesting.[45] At the time of writing (October 2014), only 18 states had deposited acceptances of the Doha Amendment to the Protocol. It is not hard to imagine that the amendment will never formally enter into force.[46]

The fact that there is a second commitment period for the Protocol running in parallel with a pledge period for the FCCC makes no sense at all from a design perspective. This highly unsatisfactory division of the regime into two tracks was a long time coming. Hopefully states will return to a single track as of 2020. From a legal point of view, the Protocol track is a continuation of the FCCC's specific

---

43 See Section 2.2.5 in Chapter 2.
44 Kyoto Protocol (2011), *Decision 1/CMP.7*.
45 Duncan French and Lavanya Rajamani (2013), 'Climate Change and International Environmental Law: Musings on a Journey to Somewhere', 25 (3) *Journal of Environmental Law* 437, p. 440.
46 See International Institute for Sustainable Development (2014), 'Summary of the Bonn Climate Change Conference: 4–15 June 2014', 12 (598) *Earth Negotiations Bulletin* 1, pp. 18–19, 34.

mitigation rule, whereas the pledges track is an exploratory attempt to develop the FCCC's general mitigation rule.

One still-developing element of the Protocol's second commitment period will be necessary for the post-2020 regime if substantive climate law is to continue to develop according to the logic of the FCCC's general mitigation rule. When the AWG-KP wrapped up without an outcome for its main agenda item, a process was envisaged to keep up the pressure on Annex I parties to make their targets more 'ambitious', even as the second commitment period got under way. Under this model, rather than having static targets for a whole period, parties would be encouraged to accept increasingly challenging emission limits with a view to reaching an aggregate reduction target of 'at least 25–40 per cent' below 1990 levels by 2020.[47] The deadline for this process was in 2014,[48] but there would be no reason in principle not to have annual or biennial (etc.) 'ambition-improvement' rounds. The process is significant because if the global emission budget is to be informed by science, the budget cannot be expected to remain fixed. An international agreement that individualizes the FCCC's general mitigation rule by fairly allocating emission limits to states with reference to a global budget must contain a mechanism to revise the targets to reflect periodic revisions in the global budget, in accordance with scientific advice.

Because 2013–2020 is a double-track period, the FCCC (and not just the Protocol) needs a process to pressure parties to make their pledges more ambitious. The Ad Hoc Working Group on the Durban Platform (ADP), established in 2011, has been tasked with closing 'the ambition gap' in the parties' pledges.[49]

4.1.5 Regime post-2020: 'Nationally determined contributions'

The FCCC parties agreed at the Durban conference in 2011 to create a new agreement for the post-2020 period outside the framework of the Kyoto Protocol.[50] This would be the fourth phase in international emission-reduction efforts.

The post-2020 regime was still very poorly developed at the time of writing in 2014. The ADP negotiations took off where the AWG-LCA left off. The latter process began in 2008. Apparently, it created nothing of value for the ADP, which started with a blank sheet. The FCCC parties have undertaken to wrap up the ADP by the end of 2015. Both the AWG-LCA and the AWG-KP had their deadlines

---

47 Kyoto Protocol (2012), *Decision 1/CMP.8*, para. 7. Art. 3.1 bis to the amended treaty (see ibid., Annex I) provides that the participating Annex I parties are to reduce their overall emissions by at least 18 per cent below 1990 levels in 2013–2020 by staying within their new assigned amounts. A revised 25–40 per cent target would represent a significant increase in ambition.
48 Ibid., Annex I, para. 9.
49 FCCC (2011), *Decision 1/CP.17, Establishment of an Ad Hoc Working Group on the Durban Platform for Enhanced Action*, para. 7.
50 FCCC (2011), *Decision 2/CP.17*.

extended twice. On the basis of past experience, the ADP will also need extra time.[51]

The key term in the post-2020 regime from a mitigation perspective is that of the 'nationally determined contribution' (NDC).[52] Like the agreement that is scheduled for the end of 2015, the new term is of 'universal' application. Distinctions between commitments and actions, targets and NAMAs, and so on, supposedly will fall away, as will the distinction between Annex I and non-Annex I parties. Distinctions of all kinds will be evident, of course, in the substance of the NDCs, but at least conceptually the single term will indicate that each party to the new agreement has a responsibility for mitigation. FCCC parties were due to submit NDCs for the new agreement in the first half of 2015.[53]

There are some indications that the new agreement will advance the essential climate law that I have described in this book. Advances in the area of accountable reporting are likely to be modest, as this area of the law is already well developed and established. Perhaps the only interesting outstanding question is whether accountable reporting should be bolstered with a compliance system that imposes penalties on states for non-compliance with reporting and mitigation obligations. The FCCC's mitigation law, on the other hand, has hardly advanced at all since the 1990s, leaving plenty to be done to develop the general mitigation rule. However, the ADP talks in the first half of 2014 were still being conducted at such an elementary level that it was impossible to tell to what extent the post-2020 agreement would advance mitigation law, if at all. On the nature of NDCs, the US position was that in order to ensure the broadest possible state participation in mitigation commitments, 'countries should be allowed to define their own mitigation contributions according to national circumstances', which the United States claimed does 'not necessarily result in a lower level of ambition' if joined to a 'consultative period' during which states have an opportunity to analyse each other's measures.[54] The US approach is of the bottom-up kind and makes

---

51 On the slow process, see International Institute for Sustainable Development (2014), 'Bonn Climate Change Conference: 10–14 March 2014', 12 (595) *Earth Negotiations Bulletin* 1, p. 4 ('Parties expressed concern over the absence of a method for developing text' for the new agreement); and ibid., p. 16 ('India, for BASIC, expressed concern with the slow progress, observing that sessions have seen further reiteration of ideas and positions already outlined at roundtables and workshops over the last two years').
52 The expression first occurs in FCCC (2013), *Decision 1/CP.19*, para. 2.b. As of 2014, NDCs were being referred to as *intended* NDCs; see International Institute for Sustainable Development (2014), 'Summary of the Bonn Climate Change Conference: 4–15 June 2014'.
53 FCCC (2013), *Decision 1/CP.19*, para. 2(b); and Ad Hoc Working Group on the Durban Platform for Enhanced Action (ADP) (4 February 2014), *Reflections on Progress Made at the Third Part of the Second Session of the Ad Hoc Working Group on the Durban Platform for Enhanced Action and on Its Work in 2014*, ADP.2014.1.InformalNote, para. 6. It is not clear what use will be made of the NDCs in the short space between the time they are submitted and the time (December 2015) when the post-2020 agreement is due to be finalized.
54 International Institute for Sustainable Development (2013), 'Summary of the Bonn Climate Change Conference: 29 April – 3 May 2013', 12 (568) *Earth Negotiations*

no reference to a global budget.⁵⁵ New Zealand is reported to have called for a 'hybrid structure of bottom-up and top-down approaches', although the latter might have been a reference to accountable reporting rather than target-setting for mitigation.⁵⁶ South Africa, on the other hand, supported 'a global goal' for mitigation which would be used to calculate 'absolute emission reduction targets for developed countries and relative targets for developing countries with deviation from business-as-usual emissions'.⁵⁷ South Africa's proposal would individualize the FCCC's general mitigation rule in a single step and represent a major advance in international climate change law.⁵⁸ However, not only does it contradict the US position, it sets the condition that developing-country targets will only be of the speculative (non-absolute) variety. Not all Annex I parties would be prepared to make this concession. South Africa also supported a compliance system for the new agreement. In this, it was backed by Switzerland, which said that such a compliance system should include 'supportive and sanctioning elements' and should adopt the institutional arrangements developed under the Protocol.⁵⁹ As

---

Bulletin 1, p. 5. See also International Institute for Sustainable Development (2013), 'Warsaw Highlights: Friday, 15 November 2013', 12 (588) *Earth Negotiations Bulletin* 1, p. 2 ('the US proposed a staged approach to maximize participation, with all parties submitting nationally determined mitigation commitments under a single but flexible set of rules applicable to all; a global consultation process; and regular reviews at the implementation stage').

55 Nevertheless, it makes reference to fairness. The United States insists that countries should explain why their NDCs are fair given their national circumstances: International Institute for Sustainable Development (2014), 'Bonn Climate Change Conference: 10–14 March 2014', p. 9.
56 International Institute for Sustainable Development (2013), 'Summary of the Bonn Climate Change Conference', p. 8. At a meeting in 2014, New Zealand said that the first round of NDCs may not be sufficient to achieve the 2°C goal and suggested inclusion of a mechanism in the agreement to increase ambition over time (International Institute for Sustainable Development (2014), 'Bonn Climate Change Conference: 10–14 March 2014', p. 9), thereby implying that the agreement would not bind parties legally to comply with the 2°C goal.
57 International Institute for Sustainable Development (2013), 'Summary of the Bonn Climate Change Conference', p. 8.
58 In addition to South Africa, Colombia, speaking for the Independent Alliance of Latin America and the Caribbean, suggested that the new agreement should contain a reference to a global mitigation goal, as well as a mechanism by which to derive NDCs from it: International Institute for Sustainable Development (2014), 'Bonn Climate Change Conference: 10–14 March 2014', p. 4. Nauru, speaking for AOSIS, called for 'quantification and assessment against the long-term global goal of keeping global warming below 1.5°C; and developing a process to review proposed national contributions': ibid., p. 3. A minority of FCCC parties would prefer to see the general mitigation rule controlled by this stricter limit.
59 International Institute for Sustainable Development (2013), 'Summary of the Bonn Climate Change Conference', p. 5. Other supporters of a compliance system included the European Union, which is reported to have stated that such a system was necessary to go 'beyond transparency': International Institute for Sustainable Development (2014), 'Bonn Climate Change Conference: 10–14 March 2014', p. 11. New Zealand and the United States, on the other hand, were content with just transparency. The latter country clarified that transparency is to be constituted of 'reporting, technical

I discussed in Chapter 3, the United States has expressed a preference for the continuation of the model of accountable reporting implemented under the FCCC in the 2013–2020 period, which does not include a compliance system.

The joint chairs of the ADP, reflecting on the direction of the negotiations in early 2014, concluded that the new agreement 'may ... also need to address the aggregate and long-term perspective in view of the upper limit of acceptable warming of 2°C agreed by Parties'.[60] However, they did not foresee a central role for this element, preferring an approach that would 'continue to raise ambition over the long term in order to achieve the objective of the Convention'.[61] Some months earlier they had mentioned the need for 'a robust process, framework or mechanism to ensure that Parties' contributions are ... equitable and fair',[62] suggesting that a burden-sharing agreement based on the comparability of state effort would form a part of, or precede, the new agreement. As we have seen, the FCCC's general mitigation rule implies that such an element is necessary for the individualization of the mitigation rule and further development of climate law, but the ADP joint chairs did not refer to it again in 2014. In relation to the accountable reporting arrangements for the FCCC's pledge period, the joint chairs said, rather vaguely: 'We may wish to consider possibilities of anchoring existing transparency arrangements in the 2015 agreement.'[63] They seemed to skirt the issue of a compliance system.[64]

In summary, by late 2014, there was still little clarity about the content of the post-2020 agreement and no sign that states would meet their obligations under the FCCC's general mitigation rule any time soon by developing it in the direction of its inherent logic.

4.2 State compliance in practice

4.2.1 Impact of Framework Convention

The FCCC's specific mitigation rule for the period 1990–2000 required Annex I parties to return to their 1990 emission levels by 2000, individually or jointly. The rule did not say whether emissions from the LULUCF sector should be included in the calculation, but the Convention's general reference to 'anthropogenic emissions' implies that LULUCF should not be left out. The FCCC Secretariat and the SBI

---

 review and a facilitated party-driven process': ibid., p. 11; in other words, the same arrangement as for the 2013–2020 period under the FCCC.
60 Ad Hoc Working Group on the Durban Platform for Enhanced Action (4 February 2014), *Reflections on Progress*, para. 8. See also International Institute for Sustainable Development (2014), 'Bonn Climate Change Conference: 10–14 March 2014', p. 9 (ADP 'Co-Chair Runge-Metzger stressed the importance of addressing mitigation at the scale required to stay below the 2°C limit of acceptable warming, and consider how to reflect this in the new agreement').
61 ADP (4 February 2014), *Reflections on Progress*, para. 9.
62 ADP (13 August 2013), *Note by the Co-Chairs: Note on Progress*, ADP.2013.14. InformalNote, p. 2.
63 ADP (4 February 2014), *Reflections on Progress*, para. 15.
64 Ibid., paras 17–18.

Table 4.3 Greenhouse gas emissions (Mt CO_2 eq.) of Annex I parties to the FCCC, including LULUCF. The shaded half-rows highlight the states that did not return their greenhouse gas emissions to 1990 levels by the year 2000 in accordance with the specific mitigation rule. Economies in transition are indicated with an asterisk (*).

Annex I party	1990	1995	2000	% change 1990–2000	2005	2010
Australia	524.0	462.9	556.4	6	534.3	587.8
Austria	68.2	68.2	65.3	−4	85.5	81.5
Belarus*	110.6	51.6	48.3	−56	57.9	59.3
Belgium	142.2	149.8	145.6	2	142.8	134.1
Bulgaria*	95.5	62.7	50.6	−47	57.5	52.2
Canada	529.5	832.9	665.4	26	800.1	804.0
Croatia*	25.3	14.0	18.6	−26	22.7	20.9
Czech Republic*	192.7	143.8	138.7	−28	140.0	134.0
Denmark	73.7	80.0	74.7	1	69.0	59.6
Estonia*	31.4	10.6	21.4	−32	9.5	16.8
Finland	55.3	56.8	48.9	−12	40.0	49.9
France	539.6	530.0	540.3	0	527.0	490.2
Germany	1,214.5	1,083.0	1,005.8	−17	1,013.1	952.2
Greece	102.1	106.1	123.5	21	132.8	114.7
Hungary*	97.0	74.7	77.8	−20	75.2	63.9
Iceland	4.7	4.4	4.9	4	4.7	5.5
Ireland	55.4	59.2	68.5	24	69.1	60.3
Italy	506.8	499.9	525.5	4	521.2	457.0
Japan	1,197.6	1,257.6	1,256.6	5	1,260.8	1,182.0
Latvia*	4.0	−9.0	−9.1	−328	−6.1	−16.4
Liechtenstein	0.2	0.2	0.2	0	0.3	0.2
Lithuania*	43.6	18.8	12.6	−71	21.0	9.8
Luxembourg	13.2	9.9	9.4	−29	12.6	12.0
Malta	1.9	2.3	2.5	32	2.9	2.9
Monaco	0.1	0.1	0.1	0	0.1	0.1
Netherlands	214.8	226.0	215.9	1	214.0	212.2
New Zealand	31.6	39.0	45.6	44	51.7	54.1
Norway	41.1	38.3	34.4	−16	23.9	21.0
Poland*	441.9	428.6	382.4	−13	353.5	383.9
Portugal	53.3	62.1	71.3	34	84.2	62.1
Romania*	230.1	156.0	112.5	−51	122.7	97.2
Russia*	3,429.8	1,966.7	1,575.7	−54	1,577.7	1,555.2
Slovakia*	61.5	42.3	39.1	−36	45.9	40.0
Slovenia*	9.4	9.6	9.0	−4	11.9	9.8
Spain	263.7	293.4	355.5	35	410.9	319.7
Sweden	35.6	42.8	33.4	−6	36.5	34.9
Switzerland	49.9	47.7	50.6	1	52.5	51.8
Turkey	173.0	218.7	252.7	46	286.0	362.0
Ukraine*	860.2	449.8	344.9	−60	378.9	345.3
United Kingdom	775.5	716.3	678.8	−12	659.9	595.3
United States	5,388.7	5,758.8	6,394.7	19	6,197.4	5,921.5
Total	17,689.2	16,066.6	16,049.0	−9	16,101.6	15,400.5
Total excluding EIT	12,056.2	12,646.4	13,226.5	10	13,233.3	12,628.6

Source: Compiled from the most recent Inventory Review Reports; see <http://unfccc.int/national_reports/annex_i_ghg_inventories/inventory_review_reports/items/6947.php>. Monaco, Slovakia, and Slovenia became Annex I parties in 1998; Malta became an Annex I party in 2010.

monitored the parties' progress in this respect. In 1994, the Secretariat wrote that 'In due time, it will be possible to assess the achievement of the [target]. At present, a comparison of CO_2 projections for 2000 with inventories for 1990, would suggest a somewhat greater need for additional measures.'[65] This comment, from early in the period controlled by the specific mitigation rule, was based on information in Annex I parties' first national communications. It suggests that the target was perceived as achievable, but not without additional effort.[66] When the second national communications were submitted, in 1997–1998, the dramatic fall in emissions from EIT countries would have been the most noteworthy new information.[67] Collective compliance with the specific mitigation rule was at that point assured.

In 2003, the SBI delivered its verdict on the 2000 target: 'The total aggregated GHG emissions ... decreased by 3 per cent from 1990 to 2000. Thus Annex I Parties have jointly attained the aim of Article 4.2 of the Convention.'[68] The SBI excluded LULUCF from its calculation, without providing a rationale for doing so.[69] Table 4.3 shows the emission totals, with LULUCF included.

The decade's Annex I emissions, including LULUCF, ended 9 per cent down on 1990 levels. The SBI acknowledged that the feat was 'mainly due' to the

---

65 FCCC Interim Secretariat (7 December 1994), *First Compilation and Synthesis of First National Communications from Annex I Parties*, A/AC.237/81, para. 14.
66 A similar caution was expressed two years later, after all national communications for the first round had been received: 'for the majority of Annex I Parties additional measures would be needed to return CO_2 emissions to their 1990 level by 2000 (FCCC Secretariat (11 June 1996), *Second Compilation and Synthesis of First National Communications from Annex I Parties: Executive Summary*, FCCC/CP/1996/12, para. 43). It is noteworthy that the Secretariat focused on individual rather than aggregate state performance, even though the Convention allows a choice. Werksman, as well as Doelle et al., interpret the specific mitigation rule as falling on states individually: Jacob Werksman (1999), 'Compliance and the Kyoto Protocol', p. 66 ('the EC, the United States, and Japan are each likely to exceed 1990 greenhouse gas emission levels in 2000 and may thus be considered to be on the path towards non-compliance with their commitments under Article 4'); and Meinhard Doelle, Dennis Mahony, and Alex Smith (2012), 'Canada', in *Climate Change Liability: Transnational Law and Practice*, ed. Richard Lord et al., p. 527 (stating that in Canada 'there has not been a serious effort nationally to meet the FCCC commitment to return to 1990 levels of emissions by 2000').
67 The main cause of the steep fall in emissions in EIT countries between 1990 and 2000 was a drastic decrease in industrial output and domestic consumption resulting from a restructuring and technological renewal of the former Soviet-bloc economies.
68 FCCC Subsidiary Body for Implementation (16 May 2003), *Compilation and Synthesis of Third National Communications from Annex I Parties: Executive Summary*, FCCC/SBI/2003/7, para. 11. The verdict was echoed in academic writings; e.g. René Lefeber (2012), 'Climate Change and State Responsibility', p. 323.
69 The likely reason for the exclusion is that the estimation of LULUCF emissions/removals is relatively uncertain and human control over them is not as pronounced as in other economic sectors. Nevertheless, Kyoto Protocol parties count certain LULUCF emissions/removals towards their assigned amount, so it is not the case that they are automatically to be excluded when the discussion turns to emission targets.

Table 4.4 Greenhouse gas emissions (Mt CO$_2$ eq.), including LULUCF, from a sample of 47 non-Annex I parties to the FCCC (about one-third of the total) arranged from highest to lowest emissions on the basis of the most recent inventory. Non-Annex I emission information, including that shown here, has not been independently reviewed, is often incomplete, and as a rule should be treated as being of relatively low reliability and accuracy. Several developing countries that claim to be net emission sinks are shown in italics at the end of the table. The assignment of emission totals to years is approximate (e.g. the 1995 column often contains 1994 data). Gaps in the columns reflect actual gaps in reporting. Asterisks (*) mark countries that had not submitted a NAMA by mid-2013.

Non-Annex I party	1990	1995	2000	2005	2010
China	–	3,650.1	–	7,045.0	–
Brazil	1,389.1	2,603.0	2,087.7	2,191.8	–
Indonesia	464.6	498.3	1,375.6	–	–
India	–	1,228.5	1,301.2	–	–
Tanzania*	129.7	952.8	–	–	–
Mexico	425.3	474.5	563.2	711.7	–
Korea (South)	263.2	–	508.3	–	–
Ecuador*	265.1	304.6	350.1	410.1	–
South Africa	330.4	361.2	–	–	–
Nigeria*	–	347.6	–	–	–
Saudi Arabia*	150.0	–	281.9	–	–
Argentina	216.3	223.3	238.7	–	–
Thailand*	–	285.8	229.1	–	–
Egypt	106.8	–	193.2	–	–
United Arab Emirates*	–	126.2	120.2	182.1	–
Colombia	129.9	152.1	177.5	179.9	–
Venezuela*	–	–	177.9	–	–
Pakistan*	–	167.1	–	–	–
Vietnam*	–	103.8	150.9	–	–
Peru	–	98.8	120.0	–	–
Bangladesh*	–	53.8	113.3	117.6	–
Jamaica*	–	116.2	–	–	–
Philippines*	–	100.7	–	–	–
Bolivia*	38.6	47.5	71.5	91.7	–
Israel	–	62.7	72.4	73.3	75.0
Morocco	–	39.9	63.3	–	–
Chile	–	45.7	43.4	59.6	–
Nepal*	–	39.3	–	–	–
Singapore	–	26.9	38.8	–	–
Ethiopia	11.7	37.9	–	–	–
Ghana	–16.8	–5.4	13.2	23.8	–
Jordan	–	18.4	20.1	–	–
Mozambique*	8.6	16.0	–	–	–
Mongolia	24.8	14.0	15.6	15.6	–
El Salvador*	–	15.7	–	14.5	–
Costa Rica	8.5	10.7	7.9	8.6	–
Dominican Republic*	7.1	14.0	7.6	–	–
Senegal*	–	–	6.4	–	–
Micronesia*	–	0.3	–	–	–
Maldives	–	0.2	–	–	–
*Rwanda**	–	–	–4.6	–2.4	–
*Kenya**	–	–6.5	–	–	–
*Guatemala**	–24.8	–	–	–	–
*Malaysia**	–	+75.6	–26.8	–	–
*Guyana**	–57.9	–56.1	–53.5	–51.6	–
Gabon	–	–494.4	–58	–	–
D.R. Congo	–	–132.2	–295.5	–132.8	–

Source: FCCC, 'Greenhouse Gas Inventory Data: Detailed Data by Party' (online database), <http://unfccc.int/di/DetailedByParty.do>.

unforeseen EIT situation.[70] If EIT countries are set aside, the period ended with a 10 per cent increase in emissions. The most accurate conclusion would seem to be that Annex I parties complied with their obligation, as a collective, but largely due to accident rather than design.

In 2010, twenty years after the FCCC's base year, fourteen Annex I parties exceeded their 1990 emission levels (see again Table 4.3). In comparison with 2000 emission levels, Annex I emissions were down, both including and excluding the EIT contribution, 'mostly as a result of the economic crisis in the period 2007–2008', as the SBI put it.[71] The global financial crisis was thus the second surprise contributor to the lowering of Annex I emissions in the two decades following 1990. A third major contributor was a relocation of production facilities to developing countries.[72]

State compliance with the other part of the FCCC's substantive law, namely the general mitigation rule, is more difficult to assess, because some aspects of it cannot be quantified. According to the analysis I have presented, the general rule divides into communal obligations—to the effect, first, that collective emissions must be kept on a trajectory leading to a stabilization of the concentration of greenhouse gases in the atmosphere at a level that avoids 2°C warming, and, second, that states must agree on a global emission budget and an equitable allocation of that budget to achieve the stabilization—as well as into two implied individualized obligations, namely that each state must engage in good-faith in negotiations towards a long-term, budget-based, solution to the climate problem and take reasonable measures to moderate its emissions in the interim.

The general mitigation rule affects Annex I and non-Annex I parties alike. We must therefore attempt to gain some understanding of the emissions from countries in the latter group over the two decades onward from 1990 (Table 4.4).

This is a patchy and probably not very reliable record. Nevertheless, the general trend suggests a growth in emissions. Only three countries (Thailand, El Salvador, and Malaysia) in the sample of 47 non-Annex I parties shown in Table 4.4 moved consistently from higher to lower emissions. By contrast, emissions doubled in China, tripled in Indonesia, and septupled in Tanzania. The moderation in Annex I-party emissions evident in Table 4.3 was more than offset by emission growth in the non-Annex I group. As noted above, part of the growth can be attributed to

---

70 FCCC Subsidiary Body for Implementation (16 May 2003), *Compilation and Synthesis of Third National Communications (Annex I)*, para. 11.
71 FCCC Subsidiary Body for Implementation (20 May 2011), *Compilation and Synthesis of Fifth National Communications from Annex I Parties: Executive Summary*, FCCC/SBI/2011/INF.1, para. 10.
72 Taking into account the emissions embodied in international trade, Caldeira and Davis write that 'In 1990, 0.4 Gt CO_2 were emitted in developing countries to subsidize consumption in developed countries. By 2008, this subsidy increased to 1.6 Gt CO_2': Ken Caldeira and Steven J. Davis (24 May 2011), 'Accounting for Carbon Dioxide Emissions: A Matter of Time', 108 (21) *PNAS* 8533, p. 8534. The underlying data for this article are from Glen P. Peters et al. (24 May 2011), 'Growth in Emission Transfers via International Trade from 1990 to 2008', 108 (21) *PNAS* 8903. For a further discussion of 'consumption-based carbon accounting', see Marco Grasso and J. Timmons Roberts (2014), 'A Compromise to Break the Climate Impasse'.

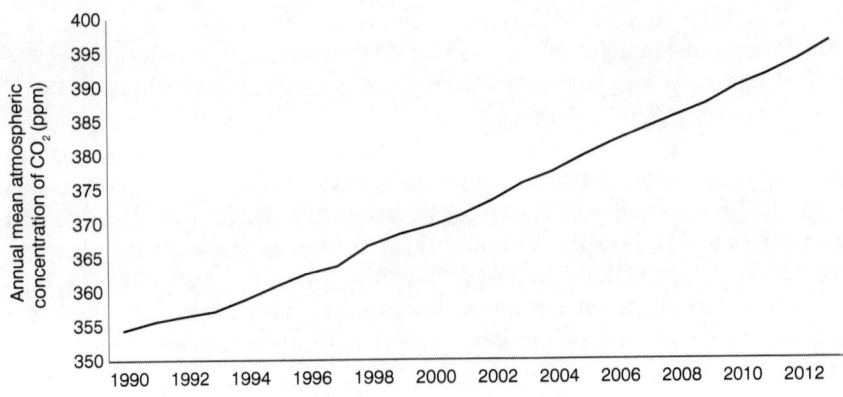

Figure 4.1 Annual mean atmospheric concentration of CO_2 in ppm since 1990, as measured at Mauna Loa.[a] By 2012, the CO_2 concentration had reached 141 per cent of the pre-industrial level.[b] (This does not take account of other greenhouse gases.[c] Counting all gases, the CO_2 eq. concentration in 2011 was 430 ppm.[d]) The concentration of CO_2 was expected to cross the 400 ppm threshold in 2015 or 2016.

Notes:
a Raw data obtained from <ftp://aftp.cmdl.noaa.gov/products/trends/co2/co2_annmean_mlo.txt>.
b World Meteorological Organization (2013), Greenhouse Gas Bulletin No. 9, p. 2.
c Atmospheric methane reached 260% of the pre-industrial level due to increased emissions from anthropogenic sources: ibid., pp. 2–3.
d IPCC (2014), Climate Change 2014: Mitigation of Climate Change: Working Group III Contribution to the Fifth Assessment Report of the Intergovernmental Panel on Climate Change: Summary for Policymakers, p. 9.

goods produced in non-Annex I parties for consumption in Annex I parties. But most of it is due to local factors.[73]

Needless to say, no specific mitigation rule applied to non-Annex I parties for the period 1990–2010. The figures shown in Table 4.4 are therefore not able to tell us anything about 'compliance' of those parties with the general mitigation rule. Most of them participated in the FCCC negotiations in good faith, although there is little evidence of non-Annex I parties supporting an individualization of the mitigation effort for any states other than the Annex I parties. On the other hand, most non-Annex I parties could make a case for having taken some measures to moderate their emission growth.

Compliance with the general mitigaton rule finally can only be assessed in general terms, yet it can also be revealing to focus on information about global emissions, the global emission budget, and the 'emission gap'. As an introduction to the first of these three, we may consider the steady rise in atmospheric CO_2 concentrations (Figure 4.1).

It is tempting to treat Mauna Loa's famous dial of CO_2 readings as the essential test of state compliance with the climate change regime. Despite all the policies,

---

73 See Marco Grasso and J. Timmons Roberts (2014), 'A Compromise to Break the Climate Impasse'.

plans, and even legislation at the domestic level, 'there has not yet been a substantial deviation in global emissions from the past trend'.[74]

In its Fifth Assessment Report, the IPCC estimated that total global anthropogenic emissions per annum from all sources, including LULUCF, rose from 38 to 40 Gt CO_2 eq. between 1990 and 2000, and from 40 to 49 Gt CO_2 eq. between 2000 and 2010,[75] the highest in human history. In the period from 1970 to 2000, the growth in emissions was 0.4 Gt CO_2 eq. (1.3 per cent) per year, whereas from 2000 to 2010 it was 1 Gt CO_2 eq. (2.2 per cent) per year.[76] In other words, emissions from FCCC parties collectively accelerated in the decade following the expiration of the specific mitigation rule in 2000.

Under a business-as-usual scenario, the global growth in emissions is expected to persist. The IPCC's baseline scenario (a future with no mitigation measures additional to those currently in place) results, by 2100, in a mean temperature rise of between 3.7 and 4.8°C above pre-industrial levels.[77] Mitigation scenarios under which it is 'likely' that the temperature rise can be kept below 2°C imply an atmospheric concentration in 2100 of about 450 ppm CO_2 eq.[78] Scenarios reaching that concentration by 2100 are characterized by 40–70 per cent lower global emissions in 2050 than in 2010, and an emission level near, or below, zero in 2100.[79]

Taking into account the amount of anthropogenic greenhouse gases already in the atmosphere, the universally agreed warming limit of 2°C, and the requirement of stabilization of CO_2 concentrations by 2100, the IPCC is able to calculate a global emission budget for compliance with the FCCC's general mitigation rule. Expressing that budget in terms of emissions of CO_2 by 2100 (the warming impact of the other greenhouse gases can be converted into a CO_2-equivalent impact and deducted from the global budget), for a probability greater than 50 per cent of remaining below 2°C compared with the period 1861–1880, cumulative CO_2 emissions from all anthropogenic sources prior to 2100 must not to exceed 4,440 Gt CO_2; and for a probability greater than 66 per cent, emissions must not exceed 3,670 Gt CO_2. An amount of 1,890 Gt CO_2 was already emitted by 2011.[80] The budget onward from 2011 was, therefore, 2,550 Gt CO_2 for a probability greater than 50 per cent, or 1,780 GtCO_2 for a probability greater than 66 per cent. At the 2010 global emission rate of 49 Gt CO_2 eq./year, the budget for the safer of the two options would be completely spent by 2047 or thereabout.

The IPCC scenario that best aligns with the specifications of the FCCC's general mitigation rule (i.e. the 2°C limit) is called RCP2.6. Relative to the pre-industrial period, warming is 'unlikely' to exceed 2°C under RCP2.6.[81] The most reasonable reading of our legal obligation under article 3 of the FCCC would lead

---

74 IPCC (2014), *5AR WG3 SPM*, p. 28.
75 Ibid., Figure SPM 1, p. 7.
76 Ibid., p. 6.
77 Ibid., p. 9.
78 Ibid., p. 10.
79 Ibid., p. 13.
80 IPCC (2013), *5AR WG1 SPM*, p. 25.
81 Ibid., p. 18.

us to select this scenario for the future. Its unspent global budget, at 990 Gt CO_2,[82] is much tighter than the last two considered. At the current rate, it would be spent by 2031. The tighter the global emission budget, the fewer the pathways for a managed, economically feasible, transition to a zero-emission world.

The third element in this analysis is the emission gap through to 2020. The reason for choosing 2020 is that current mitigation commitments extend only as far as that year (see Section 4.1.3, above). The 'gap' is the difference between the emission levels in 2020 if FCCC parties abide by their mitigation pledges, and emission levels in the same year if FCCC parties were on a pathway to avoid 2°C of warming.[83] According to the IPCC, the gap is a wide one. Emission levels in 2020 based on the pledges 'are not consistent with cost-effective long-term mitigation trajectories that are at least *as likely as not* to limit temperature change to 2°C' (i.e. they have less than a 50 per cent probability of success).[84] The pledges are instead consistent with scenarios that are 'likely' only to keep temperature change from exceeding 3°C.[85]

The FCCC parties concede the existence of an emission gap, which they also call an 'ambition' gap, and have expressed 'grave concern' about it.[86] As discussed at the end of Section 4.1.4, the FCCC parties have tasked the ADP to explore the options to close the gap.[87] From a legal point of view, none of this changes the fact that the FCCC parties, collectively, have so far failed to comply with the element of the general mitigation rule that demands convergence on a feasible pathway to avoid 'dangerous interference' with the climate system. The pathway the parties have chosen between now and 2020 all but guarantees dangerous interference.

The other communal element of the general mitigation rule is that the FCCC parties should be making progress on an agreement to individualize the mitigation burden in accordance with a global emission budget consistent with the 2°C limit. All the evidence suggests (and indeed it is intuitively obvious) that outcomes seen as equitable can lead to more effective cooperation.[88] Yet this is an area of lack of progress among FCCC state parties, and therefore of legal non-compliance. In 2013, the FCCC Secretariat reported on an project to assess 'comparability' of the mitigation efforts of state parties.[89] The step logically precedes a decision on the allocation of a global emission budget. The model put forth by the Secretariat is

---

82 Ibid., Table SPM 3, p. 25.
83 See UN Environment Programme (November 2013), *The Emissions Gap Report 2013: A UNEP Synthesis Report*, p. xi.
84 IPCC (2014), *5AR WG3 SPM*, p. 13, emphasis in original.
85 Ibid., p. 13.
86 FCCC (2012), *Decision 1/CP.18*, II.A preamble; and FCCC (2013), *Decision 1/CP.19*, p. 3 (affirming 'the significant gap between the aggregate effect of Parties' mitigation pledges [for] 2020 and aggregate emission pathways consistent with having a likely chance of holding the increase in global average temperature below 2°C').
87 FCCC (2013), *Decision 1/CP.19*, para. 1.
88 IPCC (2014), *5AR WG3 SPM*, p. 5.
89 FCCC Secretariat (2013), *Quantified Economy-Wide Emission Reduction Targets by Annex I Parties*, para. 108.

elementary, with significant gaps in its methodology, as the Secretariat itself has conceded.[90] Certainly, understanding comparability is no simple matter.[91]

No further developments on the burden-sharing project have been reported by the Secretariat since its 2013 study. We know from the ADP discussions (see Section 4.1.5) that a global budget is supported by some states but not by others. States in the latter category, which would seem to include most states, would also be reluctant to engage in systematic negotiations on effort-sharing. This whole area of legal obligation lies neglected, almost as undeveloped as it was when the FCCC went into force.[92]

The final matter to consider is whether the FCCC parties have complied with their state-level obligations entailed by the general mitigation rule, namely a good-faith engagement with the international negotiation process to individualize the rule and a reasonable (interim) mitigation effort volunteered by each state. The former question I have answered implicitly, in the negative, in the previous paragraph: most states seem content to perpetuate a practice of bottom-up mitigation commitments. The latter question, which requires a display of an appropriate mitigation stance while heavier burdens are under negotiation, is more difficult to answer. Certain countries have been repeatedly criticized for their apparent lack of commitment to their obligations under the climate change regime. Canada's climate policy, it is said, caved in to the interests of economic development and the protection of trade.[93] Russia, on one view, might as well have no domestic climate policy at all.[94] I discussed Australia's extraordinary about-turn in Chapter 1. The Obama Administration's commitment to lowering US emissions from business-as-usual levels has expressed itself only in ways that do not require

---

90 Ibid., paras 109 and 111.
91 As the FCCC Secretariat puts it, 'Different and diverse national circumstances can complicate the consideration of comparability of mitigation efforts, such as climate, geography, population, economic profile, governmental structure, natural resource endowment, transport systems, energy production and consumption patterns, and trade profile': ibid., para. 115.
92 On burden-sharing, the foundational analysis is by Lasse Ringius, Asbjørn Torvanger, and Arild Underdal (2002), 'Burden Sharing and Fairness Principles in International Climate Policy', 2 *International Environmental Agreements* 1–22. For a recent discussion of the topic, see Steffen Kallbekken, Håkon Sælen, and Arild Underdal (2014), *Equity and Spectrum of Mitigation Commitments in the 2015 Agreement*, Copenhagen, Nordic Council of Ministers. Also relevant is Friedrich Soltau (2009), *Fairness in International Climate Change Law and Policy*, Cambridge, Cambridge University Press.
93 Jane Matthews Glenn and José Otero (2013), 'Canada and the Kyoto Protocol: An Aesop Fable', in *Climate Change and the Law*, ed. Erkki J. Hollo, Kati Kulovesi, and Michael Mehling, pp. 493 and 506 ('Fossil fuel production in general, and from the tar sands in particular, is crucially important to the economy of the western provinces').
94 Yulia Yamineva (2013), 'Climate Law and Policy in Russia: A Peasant Needs Thunder to Cross Himself and Wonder', in *Climate Change and the Law*, ed. Erkki J. Hollo, Kati Kulovesi, and Michael Mehling, p. 556 (noting, however, that Russia's recent policy and legislative framework for decreasing energy intensity could deliver benefits if fully implemented: ibid., p. 561).

Table 4.5. Greenhouse gas emissions (Mt CO_2 eq.) of Annex I parties to the Protocol in the first commitment period, including Canada, which withdrew from the Protocol before the period ended. In addition to emissions from the five economic sectors listed in Annex A to the Protocol, emissions/removals from the Protocol's LULUCF activities (article 3.3 and, if elected by the party, article 3.4) are included in the totals. The 'assigned amount' for each party is shown in the second column as the annual average for the commitment period (the figure multiplied by 5 gives the total assigned amount). Shaded *rows* highlight countries tracking above their assigned amount. The shading of the 2012 *column* is to indicate that the amounts shown have not been verified through the Expert Review Team process but have been taken directly from the parties' submitted (but otherwise unprocessed) national inventory reports. All parties must include emissions/removals from Kyoto Protocol article 3.3 activities in their totals. The last column shows the parties that have elected also to count article 3.4 emissions/removals. The asterisk (*) indicates that a relatively large article 3.4 sink has been reported by the country (15 per cent or more of non-LULUCF emissions). Interstate trading of emissions is not included. Economies in transition are shown in the lower part of the table.

Annex I party	Assigned amount p/y	Emissions (first commitment period)					Notes on removals
		2008	2009	2010	2011	2012	
Australia	*591.5*	591.7	583.9	577.8	572.0	554.6	
Austria	*68.8*	85.8	78.7	83.8	81.6	78.6	
Belgium	*134.8*	137.0	125.5	135.4	120.5	116.7	
Canada	*558.4*	732.7	691.5	692.6	KP withdrawal		Art. 3.4
Denmark	*70.0*	63.2	58.3	61.4	54.3	51.1	Art. 3.4
Finland	*71.0*	35.0	19.8	43.6	35.7	27.8	*Art. 3.4
France	*563.9*	476.8	462.0	472.4	433.6	440.5	*Art. 3.4
Germany	*973.6*	942.3	878.1	910.2	883.2	889.2	Art. 3.4
Greece	*133.7*	130.0	123.3	117.0	114.7	110.0	
Iceland	*3.7*	4.4	4.2	4.0	3.8	4.1	Art. 3.4
Ireland	*56.6*	64.6	58.5	58.0	53.8	55.0	
Italy	*483.3*	507.2	453.7	461.6	458.7	430.8	Art. 3.4
Japan	*1,185.7*	1,236.6	1,160.3	1,207.9	1,255.9	1,290.2	Art. 3.4
Liechtenstein	*0.2*	0.3	0.2	0.2	0.2	0.2	
Luxembourg	*9.5*	12.3	11.8	12.3	12.1	11.7	
Monaco	*0.1*	0.1	0.1	0.1	0.1	n/a	
Netherlands	*200.3*	203.7	198.2	209.5	194.8	192.1	
New Zealand	*61.9*	57.6	54.7	54.6	56.2	60.9	
Norway	*50.1*	16.9	21.9	18.3	23.9	24.5	*Art. 3.4
Portugal	*76.4*	65.2	61.7	60.0	n/a	55.5	*Art. 3.4
Spain	*333.2*	370.4	334.9	320.4	322.0	274.3	Art. 3.4
Sweden	*75.0*	29.8	25.9	33.6	25.5	20.5	*Art. 3.4
Switzerland	*48.6*	52.5	50.5	51.6	47.4	49.4	Art. 3.4
United Kingdom	*682.4*	622.4	569.0	589.4	557.7	581.1	Art. 3.4

Annex I party – EIT	Assigned amount p/y	Emissions (first commitment period)					Notes on removals
		2008	2009	2010	2011	2012	
Belarus[a]	117.2	90.6	87.9	89.4	87.3	n/a	
Bulgaria	122.0	66.7	57.3	59.8	65.4	60.0	
Croatia	29.8	23.5	21.2	20.7	21.2	18.9	*Art. 3.4
Czech Republic	178.7	139.6	128.5	134.3	126.6	124.4	*Art. 3.4
Estonia	42.6	20.2	16.8	20.3	21.2	18.9	
Hungary	108.5	69.7	64.5	65.1	63.6	58.5	Art. 3.4
Latvia	23.8	−7.3	−6.8	−2.5	−3.3	−0.5	*Art. 3.4
Lithuania	45.5	16.9	9.4	9.3	10.7	12.3	*Art. 3.4
Poland	529.6	374.9	354.3	375.3	375.7	361.5	*Art. 3.4
Romania	256.0	127.5	102.4	100.6	102.6	96.3	*Art. 3.4
Russia	3,323.4	1,767.8	1,590.0	1,675.3	n/a	n/a	*Art. 3.4
Slovakia	66.3	49.8	44.1	45.8	44.8	41.3	
Slovenia	18.7	10.0	8.1	8.2	8.2	12.8	*Art. 3.4
Ukraine	920.8	364.8	306.7	326.9	339.7	n/a	*Art. 3.4

Note:

a An amendment to add Belarus to Annex B of the Kyoto Protocol has not yet entered into force; see <http://unfccc.int/kyoto_protocol/amendment_to_annex_b/items/4082.php>. Moreover, an Initial Review Report was not performed for Belarus.

Source: Compiled from the most recent Inventory Review Reports; see <http://unfccc.int/national_reports/annex_i_ghg_inventories/inventory_review_reports/items/6947.php>.

the support of a hostile Congress or a largely indifferent electorate.[95] The US EPA's proposed rule to limit emissions from coal-fired power stations constitutes neither a cap on emissions for the energy sector nor a shift away from fossil fuels; rather, a greater reliance on lower-emission natural gas will slow emission growth in the US energy sector, but growth itself will continue.[96] Many bleak examples may of course be drawn from the developing world. In South Africa, according to Kidd and Couzens, despite a sophisticated system of environmental laws, 'there is little (even indirect) legislation addressing climate change'.[97] A study of seventeen Latin American countries found that all of them were 'in their infancy'

---

95 Coral Davenport (2014), 'Political Rifts Slow U.S. Effort on Climate Laws', *New York Times* ('Polls consistently show that while a majority of Americans accept that climate change is real, addressing it ranks at the bottom of voters' priorities' and 'many in the Republican establishment think that talking about climate change ... would be political suicide for a Republican seeking to win the party's nomination in 2016').
96 See US EPA (2014), *Proposed Rule on Carbon Pollution Emission Guidelines for Existing Stationary Sources: Electric Utility Generating Units*.
97 Michael Kidd and Ed Couzens (2013), 'Climate Change Responses in South Africa', in *Climate Change and the Law*, ed. Erkki J. Hollo, Kati Kulovesi, and Michael Mehling, p. 620.

as far as climate law and policy goes.⁹⁸ Argentina, among them, presented a weak institutional framework for climate change and had no budgetary allocation for addressing it; as for the government's promotion of renewable-energy sources, this was attributable mainly to energy-security concerns.⁹⁹ In India, as in many developing countries, strong economic growth and poverty alleviation overwhelm other interests. India's emissions are projected to grow by around 50 per cent over the 2013–2020 period.¹⁰⁰ The country seeks to attain a grander role on the world stage, yet its climate change plans are limited to energy-conservation schemes.¹⁰¹

4.2.2 Assessment of Kyoto Protocol period I

State compliance with the emission limits for the Kyoto Protocol's first commitment period is the easiest aspect of the climate change regime to assess. Table 4.5 summarizes the information on emissions from that period.

Reflecting on the performance of Annex I parties to the Protocol during the first commitment period, several elements are noteworthy. First, most states stayed within their assigned amount without any apparent difficulty. Six states exceeded their assigned amount by a small margin. They will easily make up the difference by acquiring emission allowances through the flexibility mechanisms. Second, Canada's case stands out as highly exceptional. The country dramatically underestimated its emission needs for the period or did not make a serious effort to manage its emissions (or both). Third, the assigned amounts for EITs were hugely overestimated.¹⁰² As a result, the Protocol itself cannot have provided much of an incentive to EIT parties to implement long-term mitigation measures. In retrospect, early enthusiasm about the Protocol's targets being 'the most ambitious environmental commitments ever set by an international agreement'¹⁰³ seems misplaced. Fourth, more than a third of the parties achieved relatively large emission reductions by relying on article 3.4 sink activities. These are, of course, much more uncertain than reductions in non-LULUCF sectors.

While compliance with the emission limits of the first commitment period could be said to be high overall, in a broader context the success is only a technical one.

---

98 Soledad Aguilar and Eugenia Recio (2013), 'Climate Law in Latin American Countries', in *Climate Change and the Law*, ed. Erkki J. Hollo, Kati Kulovesi, and Michael Mehling, p. 653.
99 Ibid., p. 675.
100 Patodia Rastogi (2013), 'India's Evolving Climate Change Strategy', in *Climate Change and the Law*, ed. Erkki J. Hollo, Kati Kulovesi, and Michael Mehling, pp. 606–607.
101 Ibid., p. 613. Energy-efficiency policies that are simply business-as-usual are not necessarily policies that amount to evidence a mitigation effort. For example, the IPCC finds that 'Direct CO_2 emissions from the energy-supply sector are projected to almost double or even triple by 2050 compared to the level of 14.4 Gt CO_2 per year in 2010, unless energy intensity improvements can be significantly accelerated beyond the historical development': IPCC (2014), *5AR WG3 SPM*, p. 21.
102 For example, already at the time of the adoption of Kyoto Protocol in 1997, Russia's emissions were about 50 per cent lower than in 1990 (see Table 4.3).
103 Jacob Werksman (1999), 'Compliance and the Kyoto Protocol', p. 49.

Because the Protocol's rule is a reincarnation of the FCCC's specific mitigation rule (see Section 4.1.2), even complete compliance with the 2008–2012 emission limits would represent insignificant progress on compliance with the general mitigation rule. The Protocol's emission limits have no meaningful conceptual relationship with the 2°C threshold.

4.3 Conclusion: Limited development of the general mitigation rule

The invisibility of the FCCC's general mitigation rule has led some commentators to assume that states are currently operating in a climate law vacuum.[104]

However, the rule represents a complex of legal obligation that is undoubtedly binding on states. Implementation of the FCCC's general mitigation rule has been held back in part because of the priority given to the implementation of the Kyoto Protocol's first commitment period. Legally, there is no justification for the neglect of the general rule. The Annex I/non-Annex I distinction has long ago ceased to be a sensible basis on which to decide responsibility for mitigation. As Tables 4.6 and 4.7 at the end of the chapter show, per-capita emissions in Annex I and non-Annex I parties are no longer always far apart. Moreover, significant population growth and consequent emission increases[105] are expected in non-Annex I parties, which is not the case for the Annex I group.

The 2013–2020 pledge period under the FCCC evidences a small shift in favour of the implementation of the general mitigation rule. However, it should be emphasized that most non-Annex I parties did not submit pledges for the period. Of those that did, most did not commit even to a business-as-usual reduction. Twelve chose types that allow for their emissions to continue to rise in absolute terms. The four that chose absolute reductions currently have insignificant emission levels. Moreover, almost all non-Annex I parties have conditioned their actions on the provision of financial support. As Brunnée has argued, 'in the absence of shared understandings on burden-sharing, it will be difficult to arrive at a legitimate global commitment regime'.[106]

The ADP has so far not contributed to the development of the general mitigation rule. By mid-2014, it had produced what it modestly described as a 'landscape of issues'.[107] The landscape consisted of little more than a collection of points summarizing different and often contradictory party preferences.[108]

---

104 For example, Jutta Brunnée (2012), 'Promoting Compliance with Multilateral Environmental Agreements', in *Promoting Compliance in an Evolving Climate Regime*, ed. Jutta Brunnée, Meinhard Doelle, and Lavanya Rajamani, p. 52; and Duncan French and Lavanya Rajamani (2013), 'Climate Change and International Environmental Law', p. 440.
105 IPCC (2014), *5AR WG3 SPM*, p. 8 ('economic and population growth continue to be the most important drivers of increases in CO_2 emissions from fossil fuel combustion').
106 Jutta Brunnée (2012), 'Promoting Compliance', p. 49.
107 ADP (17 April 2014), *Reflections on Progress Made at the Fourth Part of the Second Session of the Ad Hoc Working Group on the Durban Platform for Enhanced Action*, ADP.2014.3.InformalNote, p. 5.
108 See ibid., pp. 6–15.

Table 4.6 Population 2000–2030 (thousands) and per-capita emissions (in tonnes CO_2 eq., including LULUCF) for the years 2000 and 2010, in twelve Annex I parties to the FCCC. Compare the per-capita emissions in this table with those in Table 4.7, and note, for example, that Brazil's are equal to Germany's and higher than Japan's.

Annex I		2000	2010	2020	2030
Australia		19,259	22,404	25,440	28,336
	per cap.	*28.9*	*26.2*		
Canada		30,697	34,126	37,612	40,617
	per cap.	*21.7*	*23.6*		
France		59,213	63,231	66,570	69,286
	per cap.	*9.1*	*7.8*		
Germany		83,512	83,017	81,881	79,552
	per cap.	*12.0*	*11.5*		
Italy		56,986	60,509	61,386	61,212
	per cap.	*9.2*	*7.6*		
Japan		125,715	127,353	125,382	120,625
	per cap.	*10.0*	*9.3*		
New Zealand		3,858	4,368	4,814	5,208
	per cap.	*11.8*	*12.4*		
Poland		38,351	38,199	38,158	37,448
	per cap.	*1.9*	*1.6*		
Russia		146,763	143,618	140,011	133,556
	per cap.	*10.7*	*10.8*		
Spain		40,283	46,182	47,789	48,235
	per cap.	*8.8*	*6.9*		
United Kingdom		58,951	62,066	65,600	68,631
	per cap.	*11.5*	*9.6*		
United States		284,594	312,247	337,983	362,629
	per cap.	*22.5*	*19.0*		

Source: Population figures are from United Nations Department of Economic and Social Affairs (2013), *World Population Prospects: The 2012 Revision. Volume I: Comprehensive Tables*, ST/ESA/SER.A/336, pp. 108–118; emission amounts are calculated using information from Table 4.3 in the present chapter.

Table 4.7 Population 2000–2030 (thousands) and per-capita emissions (in tonnes of CO_2 eq., including LULUCF) for the years 2000 and 2010 (subject to availability), in twelve non-Annex I parties to the FCCC. Note the significant population growth in most of these countries compared with that shown in Table 4.6.

Non-Annex I		2000	2010	2020	2030
Argentina		36,903	40,374	43,835	46,859
	per cap.	*6.5*	–		
Bangladesh		132,383	151,125	169,566	185,064
	per cap.	*0.9*	–		
Brazil[a]		174,505	195,210	211,102	222,748
	per cap.	*12.0*	*11.2*		
China[b]		1,287,696	1,367,406	1,441,044	1,461,884
	per cap.	*2.8*	*5.2*		
Egypt		66,137	78,076	91,062	102,553
	per cap.	*3.0*	–		
India		1,042,262	1,205,625	1,353,305	1,476,378
	per cap.	*1.2*	–		
Indonesia		208,939	240,676	269,413	293,482
	per cap.	*6.6*	–		
Korea (South)		45,977	48,454	50,769	52,190
	per cap.	*11.1*	–		
Mexico[a]		103,874	117,886	131,955	143,663
	per cap.	*5.4*	*6.0*		
Nigeria[c]		122,877	159,708	210,159	273,120
	per cap.	*2.8*	–		
South Africa[c]		44,846	51,452	55,131	58,096
	per cap.	*8.1*	–		
Vietnam		80,888	89,047	97,057	101,830
	per cap.	*1.9*	–		

Notes:
[a] Per-capita emissions shown are for 2000 and 2005.
[b] Per-capita emissions shown are for 1995 and 2005.
[c] Per-capita emissions shown are for 1995.

Source: Population figures are from United Nations Department of Economic and Social Affairs (2013), *World Population Prospects: The 2012 Revision. Volume I: Comprehensive Tables*, ST/ESA/SER.A/336, pp. 108–118; emission amounts are calculated using information from Table 4.4 in the present chapter.

5 Financial support for mitigation actions in developing countries

5.1 Introduction: Finance for mitigation

An agreement to attenuate climate change cannot be successful without the transfer of money and know-how from developed to developing countries to support low-emission development and adaptation to climate change. Emission reduction, many forms of which are considered too expensive even in developed countries, is often unaffordable in countries in the least-developed group. The practical reality of the situation is clear: a concerted global action against climate change requires the transfer of resources from wealthier to poorer countries. The opportunities in the mitigation area are also clear: it is often cheaper to avoid an increase in emissions in the FCCC's non-Annex I parties than to reduce emissions by the same amount in Annex I parties. Whilst these facts are evident, the applicable law and the extent of state compliance with it are less clear.

As detailed in the next section, Annex I parties have assumed legal obligations under the FCCC to provide financial support for mitigation and adaptation in developing countries. This advances the FCCC's general mitigation rule, whose objective is to have states cooperate to avoid dangerous climate change. The rule on the transfer of finance to non-Annex I parties could be treated as a separate legal provision of international climate change law dealing specifically with financial support. Alternatively, its two streams may be separated into adaptation and mitigation funding, with the latter understood as a special case of the FCCC's general mitigation rule.

The second interpretation is not as intuitive, but it is more convincing. Mitigation finance is a means to an end, and its end is the same as the FCCC's general mitigation rule. Emission reductions by non-Annex I parties funded by Annex I parties are equivalent to Annex I parties engaging in mitigation themselves (as the Protocol recognizes through its Clean Development Mechanism). The subordination of mitigation finance to the general mitigation rule can be demonstrated through the following analysis. First, global emissions in a business-as-usual scenario are estimated, and the most cost-efficient mitigation measures are planned to achieve emission reductions consistent with a 2°C pathway. Some of the measures will be implemented in developed countries and others in developing countries. Second, the total cost of the measures to be implemented in developing countries is estimated and is reduced by an amount that developing countries can themselves afford to expend on mitigation in accordance with their international obligations (i.e. the general mitigation rule). The difference in the amount represents the global transfer budget for mitigation.[1]

Financial assistance must also be supplied by Annex I parties to offset, in full or in part, the cost of compliance of developing countries with reporting obligations whose expense they cannot afford. To avoid multiplying laws unnecessarily, one might consider bringing this obligation under the rule on accountable reporting (see Chapter 2). However, the transparency rule applies to countries individually, so it is difficult to see how it could give rise to an obligation in a *group* of states to provide financial support to another group. It is therefore again preferable to conceptualize the obligation on Annex I parties to provide finance for reporting purposes to non-Annex I parties as an aspect of the general mitigation rule. This is because implementation of the rule would be rendered ineffective where there is no information, or only poor-quality information being produced at irregular intervals, on the mitigation actions and emissions of non-Annex I parties.[2] When finance is received by non-Annex I parties, for substantive actions as well as for reporting on them, the recipient country must comply with the rules on accountable reporting. I concede that this interpretation is not entirely satisfactory. It tends to collapse the accountable reporting rule into the general mitigation rule, instead of keeping them as separate procedural and substantive rules. On the positive side, the murky area of states' climate-finance obligations can be analysed more easily if it is subordinated to the rule with which it has the greatest affinity, namely the FCCC's general mitigation rule.

In the previous chapter, I argued that the logic of the general mitigation rule requires states to work in good faith towards a burden-sharing agreement through which to individualize the general rule. With the obligation to financially support poorer states understood as an aspect of the general mitigaton rule, a burden-sharing agreement is needed to individualize both emission-reduction and mitigation-finance obligations.

The conceptual separation of financial support for mitigation from that for adaptation may strike some readers as counterintuitive. Certainly, the political rhetoric at FCCC meetings insistently keeps the two united. In that narrative, climate law has settled upon a general obligation, undifferentiated as between mitigation and adaptation, to the effect that wealthy countries must fi0nancially assist poorer countries to counteract climate change. Yet, from a legal point of view, two problems arise with this idea: First, adaptation has a different rationale from mitigation. Adaptation measures (e.g. desalination plants or defences against sea incursion) do not necessarily reduce emissions. Second, finance for mitigation can in theory be quantified once a global emission budget is agreed

---

1 See Susanne Olbrisch et al. (2011), 'Estimates of Incremental Investment for and Cost of Mitigation Measures in Developing Countries', 11 (3) *Climate Policy* 970, p. 972.
2 Oberthür considers the Kyoto Protocol's reporting obligations to be contained within the mitigation obligation: Sebastian Oberthür (2014), 'Options for a Compliance Mechanism in a 2015 Climate Agreement', 4 (1–2) *Climate Law* 30, p. 36.

upon, whereas there is no objective way to agree on finance for adaptation as there is no equivalent to the 2°C target in the adaptation area.³ How, then, could one and the same law cover two such different things? The FCCC's rhetoric on non-differentiation can be explained as a political reaction to the fact that most climate finance to date has gone to mitigation projects, and many states now want financial support to be better 'balanced'.

In addition to the above, and as discussed in Chapter 1, the FCCC does not contain anything that qualifies as 'adaptation law', let alone a law of adaptation finance.⁴ There is no specific normative law for adaptation to climate change, nor is it clear that such a notion makes sense. (As Nagle says, 'If the climate was changing naturally, then we would have to try to adapt to it.'⁵) Adaptation is defensive, while mitigation is about desisting from action that is destructive. There is a clear imperative in international law to mitigate collective emissions to prevent further destruction. The need to adapt increases in proportion to state non-compliance with the imperative to reduce emissions.⁶ Climate law's affinity

---

3 Proposals for 'a global goal for adaptation' have been made before the ADP (4 February 2014), *Reflections on Progress Made at the Third Part of the Second Session of the Ad Hoc Working Group on the Durban Platform for Enhanced Action and on Its Work in 2014*, ADP.2014.1.InformalNote, para. 9), but the idea remains ill-defined and controversial (see International Institute for Sustainable Development (2014), 'Bonn Climate Change Conference: 10–14 March 2014', 12 (595) *Earth Negotiations Bulletin* 1, pp. 5–6). See also IPCC (2014), *Climate Change 2014: Impacts, Adaptation, and Vulnerability: Working Group II Contribution to the Fifth Assessment Report of the Intergovernmental Panel on Climate Change: Summary for Policymakers*, p. 28 ('Limited evidence indicates a gap between global adaptation needs and the funds available for adaptation (*medium confidence*). There is a need for a better assessment of global adaptation costs, funding, and investment. Studies estimating the global cost of adaptation are characterized by shortcomings in data, methods, and coverage (*high confidence*)'). See also the critiques in Joel B. Smith et al. (2011), 'Development and Climate Change Adaptation Funding: Coordination and Integration', 11 (3) *Climate Policy* 987, p. 988 ('there is no operational definition that can be used to specify adaptation measures and then estimate the associated capital and operating costs'); and Urvashi Narain, Sergio Margulis, and Timothy Essam (2011), 'Estimating Costs of Adaptation to Climate Change', 11 (3) *Climate Policy* 1001, p. 1011 ('current studies on adaptation costs suffer from severe limitations').

4 See J. B. Ruhl and James Salzman (2013), 'Climate Change Meets the Law of the Horse', 62 (4) *Duke Law Journal* 975, where it is argued that climate adaptation law is, substantively, like 'horse law'; see also Chris Hilson (2013), 'It's All About Climate Change, Stupid! Exploring the Relationship between Environmental Law and Climate Law', 25 (3) *Journal of Environmental Law* 359, p. 369 ('climate adaptation law ... is really covered by the sub-fields of environmental law ... such as water law and biodiversity law').

5 John Copeland Nagle (2010), 'Climate Exceptionalism', 40 *Environmental Law* 53, p. 83.

6 See IPCC (2014), *5AR WG2 SPM*, p. 28 ('Prospects for climate-resilient pathways for sustainable development are related fundamentally to what the world accomplishes with climate-change mitigation (*high confidence*). Since mitigation reduces the rate as well as the magnitude of warming, it also increases the time available for adaptation to a particular level of climate change, potentially by several decades. Delaying mitigation actions may reduce options for climate-resilient pathways in the future').

should in the first place be to the proximate cause of the problem. States must of course continue to provide relief and development assistance to those countries in need. Most likely, a lot more of it will become necessary as the impacts of climate change worsen. But the general obligation to assist developing countries to prepare for, or respond to damage from natural forces (i.e. adapt) is not a novel one. It long predates the FCCC.[7]

These introductory remarks are by way of an explanation of why in this chapter I pursue a focus on the law of mitigation finance, as distinct from climate finance in general.

In the area of 'transfer' law there is also the concept of 'technology transfer'. Transfer of technology could be for the purpose of mitigation, adaptation, or both. It is not easily quantified, in contrast with mitigation finance. Technology transfer is best thought of as the principle that Annex I technologies necessary for mitigation and adaptation should be made more accessible to developing countries. It can be derived from the law of mitigation finance and probably also from other elements of international law. While it comes up a lot in the sources on mitigation finance, I do not separate it out for special treatment in this chapter.

5.2 State obligations and compliance on finance

The FCCC calls on developed countries, and in particular on Annex II parties (the wealthier countries in the Annex I group), to assist developing countries by means of finance and technology transfer.

Article 4.3 of the Convention states that Annex I parties

> shall provide new and additional financial resources to meet the agreed full costs incurred by developing country Parties in complying with their obligations under Article 12, paragraph 1 [communication of information relating to implementation]. They shall also provide such financial resources, including for the transfer of technology, needed by the developing country Parties to meet the agreed full incremental costs of implementing measures that are covered by paragraph 1 of this Article [the commitment by developed and developing countries alike to implement policies and take action on mitigation and adaptation] and that are agreed between a developing country Party and the international entity or entities referred to in Article 11 [on the financial mechanism].

This article creates a legal obligation to financially support mitigation (and reporting on mitigation), not by relabeling existing aid but through additional

---

7 Consider in this connection the following comment by the Norwegian government, to the effect that climate adaptation funding is no more than an extension of development aid: 'It is ... very difficult to single out assistance for climate change adaptation from more general development assistance, which contributes to making countries more resilient towards climate change impacts.' Government of Norway (2013), *Norwegian Climate Finance 2012*, p. 8.

effort.[8] There is also a requirement that an overall transfer amount be agreed to first, based on needs. Thus the concepts of a global emission budget (see Chapter 4) and a global financial-transfer budget are both present in the FCCC in an elementary form.

Article 4.3, in its continuation, requires the financial flows in support of developing countries to be adequate and predictable (Annex I parties 'shall take into account the need for' these features). The provision mentions 'the importance of appropriate burden-sharing' among Annex I parties in the discharge of the article 4.3 obligations. While the text does not explain how the burden of the overall transfer amount would be shared among Annex I parties, some kind of burden-sharing agreement is recognized as necessary to give effect to the provision. It mirrors the requirement of a burden-sharing agreement for emission reductions which is an element of the FCCC's general mitigation rule.

The form of the rule is communal, so it does not bind individual states to specific payments, but it does require all states to work in good faith towards realizing such an outcome. The fact that article 4.3 mentions financial support for reporting in the same context as support for mitigation and adaptation is consistent with the argument I have developed in this book that a foundational rule of international climate change law is that all states must report their emissions transparently and fully.

Next, article 4.4 of the FCCC directs Annex I parties to assist non-Annex I parties 'that are particularly vulnerable to the adverse effects of climate change in meeting costs of adaptation to those adverse effects'. It is only an elaboration of article 4.3, emphasizing that adaptation finance must be provided to Least Developed Countries.

We come lastly to article 4.5, according to which Annex I parties

> shall take all practicable steps to promote, facilitate and finance, as appropriate, the transfer of, or access to, environmentally sound technologies and know-how to other Parties, particularly developing country Parties, to enable them to implement the provisions of the Convention. In this process, the developed country Parties shall support the development and enhancement of endogenous capacities and technologies of developing country Parties.

The FCCC assigned to the SBSTA the task of advising the Conference of the Parties 'on the ways and means of promoting development and/or transferring' the relevant technologies to non-Annex I countries.[9] From a legal perspective, article 4.5 does no more than reinforce the legal rule established through article 4.3.

Since parties individually have obligations under the accountable reporting rule (Chapter 2), and parties collectively are subject to the general mitigation rule (Chapter 4), what is the nature of the obligations under the financial-transfer

---

8 For the meaning of the term 'new and additional', see Farhana Yamin and Joanna Depledge (2004), *The International Climate Change Regime: A Guide to Rules, Institutions and Procedures*, pp. 276–278.
9 FCCC, art. 9.2.c.

rule, considering that it creates two groups of parties (transferors and recipients)? Article 4.7 of the Convention appears to address this question:

> The extent to which developing country Parties will effectively implement their commitments under the Convention will depend on the effective implementation by developed country Parties of their commitments under the Convention related to financial resources and transfer of technology and will take fully into account that economic and social development and poverty eradication are the first and overriding priorities of the developing country Parties.[10]

It might be thought that article 4.7 renders the obligations of *non*-Annex I parties 'conditional', and therefore that they are not fully legal obligations. However, this article conditions the 'effective implementation' of the commitments of non-Annex I parties—not the commitments themselves. While the article does not state that each party must negotiate in good faith to individualize obligations under the transfer rule, it is implicit in the logic of the communal rule. A non-Annex I party, or group of them, could hold out unreasonably for a higher overall transfer budget, and in the meantime fail to effectively implement their commitments under the Convention due to insufficient funds. This would in no way absolve them of their mitigation obligations.[11]

In articles 11 and 21.3, the FCCC provides that financial transfers are to be administered by a financial mechanism accountable to the Conference of the Parties. There is no detail in the Convention itself on what the mechanism would be or how it would work. The institutions needed to facilitate compliance with the financial-transfer obligation were cobbled together from the material that was at hand. It took almost two decades for the parties to establish dedicated institutions for that role.

The Global Environment Facility, which was at hand at the time the FCCC was opened for signature, was restructured to play an 'interim' role in the Convention's financial mechanism. It still continues in that role, even as the new Green Climate Fund is being set up (on the GCF, see further below). Annex I parties replenish the GEF at regular intervals. The Conference of the Parties has resolved that, prior to each GEF replenishment, the COP would make an assessment of the amount necessary to assist developing countries; however, 'These assessments have not ... led to concrete numbers being agreed by the COP'.[12] Until the Copenhagen

---

10 Ibid., art. 4.7.
11 See Jutta Brunnée and Stephen J. Toope (2010), *Legitimacy and Legality in International Law: An Interactional Account*, p. 162 ('the factual statement about implementation [does not] make the underlying developing country obligations conditional on assistance').
12 Yulia Yamineva and Kati Kulovesi (2013), 'The New Framework for Climate Finance under the United Nations Framework Convention on Climate Change: A Breakthrough or an Empty Promise?', in *Climate Change and the Law*, ed. Erkki J. Hollo, Kati Kulovesi, and Michael Mehling, p. 196.

Accord in 2009 (see below), there was never an agreed quantification of the overall financial support required under article 4.3 of the FCCC.

The GEF administers the GEF Trust Fund, the Least Developed Countries Fund, and the Special Climate Change Fund.[13] The LDCF and SCCF were established at the request of the FCCC, and are thus subject to the 'guidance' of the COP.[14] The LDCF aims to meet the needs of LDCs that are especially vulnerable to the impacts of climate change—thus linking with article 4.4 of the Convention, quoted above. It provides financial support for the preparation of National Adaptation Programs of Action. The SCCF's priorities are adaptation, technology transfer, and associated capacity-building activities.[15] Neither of the special funds, it will be noted, prioritizes mitigation actions in developing countries.

The financial-transfer rule was not further developed through the Kyoto Protocol. In its article 11, the Protocol affirms the obligation of Annex I parties to provide such financial resources as are needed by developing countries to meet the 'agreed full costs' of the implementation of commitments already established under article 4 of the FCCC.[16] The Protocol did contribute a new fund, the Adaptation Fund,[17] which of course is limited to adaptation. It receives its income mainly from a tax on the sale of flexibility-mechanism emission allowances.[18] Here we should note that mitigation mechanisms—for this is what the flexibility mechanisms are—are being used by the international regime to fund adaptation

---

13 FCCC (2010), *Decision 2/CP.16, Fourth Review of the Financial Mechanism*, FCCC/CP/2010/7/Add.2, para. 5.
14 The politics behind this measure is laid out by Yamineva and Kulovesi: 'Developing countries ... objected to the GEF because of the lack of transparency over its workings and a governance structure dominated by donor governments which was in "sharp contrast" to consensus-based decision-making procedures under MEAs, including the FCCC': Yulia Yamineva and Kati Kulovesi (2013), 'Framework for Climate Finance', p. 200.
15 Farhana Yamin and Joanna Depledge (2004), *International Climate Change Regime*, p. 290.
16 Kyoto Protocol, art. 11.2.b. See also ibid., art. 10.c, on technology transfer and intellectual property rights.
17 The Adaptation Fund was established by FCCC (2001), *Decision 10/CP.7, Funding under the Kyoto Protocol*, FCCC/CP/2001/13/Add.1. The fund's institutional arrangements were established by Kyoto Protocol (2007), *Decision 1/CMP.3, Adaptation Fund*, FCCC/KP/CMP/2007/9/Add.1. On the fund, see Britta Horstmann and Achala Chandani Abeysinghe (2011), 'The Adaptation Fund of the Kyoto Protocol: A Model for Financing Adaptation to Climate Change?', 2 (3) *Climate Law* 415.
18 The CDM's contribution amounts to 2 per cent of CERs issued for projects in non-LDC countries: Kyoto Protocol (2005), *Decision 3/CMP.1, Modalities and Procedures for a Clean Development Mechanism as Defined in Article 12 of the Kyoto Protocol*, FCCC/KP/CMP/2005/8/Add.1, Annex, para. 66(a). The tax on AAU and ERU sales is a more recent innovation: 'for the second commitment period, the Adaptation Fund shall be further augmented through a 2 per cent share of the proceeds levied on the first international transfers of AAUs and the issuance of ERUs for Article 6 projects.' Kyoto Protocol (2012), *Decision 1/CMP.8, Amendment to the Kyoto Protocol Pursuant to Its Article 3, Paragraph 9 (Doha Amendment)*, FCCC/KP/CMP/2012/13/Add.1, Annex I, para. 21.

instead of more mitigation. The Adaptation Fund along with the CDM are institutions of financial transfer that are fully controlled by the climate change regime, and hence are in a different category than the GEF. The Protocol, like the FCCC, is silent on the level of resources to be supplied by Annex I parties for transfer purposes.

The next finance-related development was the FCCC's Bali Action Plan of 2007. Under that plan, financial contributions from Annex I parties for mitigation measures in developing countries were to be guided by Nationally Appropriate Mitigation Actions.[19] Progress towards the plan's implementation became evident only at the December 2010 Conference of the Parties in Cancun, in the context of negotiations on the FCCC's future. The Cancun conference called yet again on developed countries to provide developing countries with new, additional, and long-term finance to implement plans and concrete actions for mitigation and adaptation.[20] A quantified promise of new funds for developing countries that had appeared in the Copenhagen Accord in 2009 was elevated by the Cancun conference to a decision of the FCCC. The amounts were given in arbitrary round figures, yet for the first time the commitment under the rule on financial support was expressed in dollar amounts.

The Copenhagen Accord promised different amounts for two periods, and this was carried over into the FCCC decision. For the period 2010–2012, the support would be referred to as fast-start finance and would amount to $30 billion in total.[21] For the subsequent period, 2013–2020, which is the same as the FCCC's 'pledge' period for emission reductions, Annex I parties pledged to gradually increase their support so that it would reach $100 billion per year by 2020. The new money was to be split equally between mitigation and adaptation, which if followed through would of course mean that mitigation finance would rise to only $50 billion annually by the end of the decade. Not all of the finance was to come from the public sector; some of it would be 'mobilized' from private and 'alternative' sources.[22] The promised support was conditioned on 'meaningful mitigation actions and transparency on implementation' by developing countries.[23] The FCCC set up a registry to record NAMA projects seeking international finance and match them with committed funds, thus creating a link between mitigation finance and the emission-reduction pledges of non-Annex I parties for 2013–2020.[24]

---

19 FCCC (2007), *Decision 1/CP.13, Bali Action Plan*, FCCC/CP/2007/6/Add.1, para. 1. NAMAs were discussed in Chapter 2.
20 FCCC (2010), *Decision 1/CP.16, The Cancun Agreements: Outcome of the Work of the Ad Hoc Working Group on Long-Term Cooperative Action under the Convention*, FCCC/CP/2010/7/Add.1, para. 18.
21 Ibid., para. 95; FCCC (2011), *Decision 3/CP.17, Launching the Green Climate Fund*, FCCC/CP/2011/9/Add.1, Annex, para. 3.
22 FCCC (2010), *Decision 1/CP.16*, para. 99.
23 Ibid., para. 98.
24 Ibid., paras 54–56. NAMAs in existence in 2010 were compiled into a document by the Secretariat: FCCC Secretariat (18 March 2011), *Compilation of Information on Nationally Appropriate Mitigation Actions to Be Implemented by Parties Not Included in Annex I to the Convention*, FCCC/AWGLCA/2011/INF.1. Registration of NAMAs

There was no agreement on burden-sharing in relation to the $30 billion or the funding for the pledge period, so the obligation to provide the above sums retained its 'communal' form.[25] It was left to the initiative (bottom-up action) of states to determine who was to give what.

A 2007–2008 study by the FCCC Secretariat presented an assessment of the investment and financial flows that would be needed for mitigation efforts through to 2030 consistent with a non-dangerous stabilization of CO_2 concentrations.[26] Combining several estimates of the mitigation potential of different sectors, the Secretariat estimated that to reduce emissions in 2030, projected to stand at 61.5 Gt CO_2 eq. under business as usual (including LULUCF), to 29.1 Gt CO_2 eq. (equivalent to 25 per cent below 2000 levels), climate finance would need to rise to $380 billion per annum by 2030, of which $176 billion in that year would be required for mitigation in developing countries.[27] At around the time of the FCCC's study, the IEA analysed options for reducing energy-related CO_2 emissions by 2030. These emissions were projected to rise under business as usual to 40.2 Gt CO_2 eq. in 2030.[28] In the IEA's '450 ppm stabilization scenario', global emissions must fall to 37.1 Gt CO_2 eq. in 2030, approximately equalling 1990 emissions, and for this to happen, energy-related CO_2 emissions must be reduced by about 35 per cent from business as usual.[29] The average global investment for 2021–2030 would then have to be $808 billion per annum, of which $350 billion per annum would be required in developing countries.[30]

The estimates reviewed above differ considerably from each other. They also differ from the figure committed to by FCCC parties, of $100 billion in 2020, of which only $50 billion is for mitigation (if half the total in fact goes to adaptation). Mitigation finance for 2010–2020 was clearly decided arbitrarily. There was little

---

and donor funding commenced in 2012: FCCC (2011), *Decision 2/CP.17, Outcome of the Work of the Ad Hoc Working Group on Long-Term Cooperative Action under the Convention*, FCCC/CP/2011/9/Add.1, paras 46–48. The first 'match' recorded by the registry occurred in May 2014: see <http://climate-l.iisd.org/news/nama-registry-records-first-match/>. See also Yulia Yamineva and Kati Kulovesi (2013), 'Framework for Climate Finance', pp. 220–221, expressing doubts about the added value of the NAMA registry given the Green Climate Fund's window for mitigation finance.

25 Lefeber calls it 'a collective commitment': René Lefeber (2012), 'Climate Change and State Responsibility', in *International Law in the Era of Climate Change*, ed. Rosemary Rayfuse and Shirley V. Scott, p. 326.
26 FCCC Secretariat (2007), *Investment and Financial Flows to Address Climate Change*, p. 24.
27 FCCC Secretariat (2008), *Investment and Financial Flows to Address Climate Change: An Update*, FCCC/TP/2008/7, paras 59–60 and 242. Another FCCC report from the period follows no particular methodology, mainly relying on funding requests made by non-Annex I parties: FCCC Subsidiary Body for Implementation (14 November 2007), *An Assessment of the Funding Necessary to Assist Developing Countries in Meeting Their Commitments Relating to the Global Environment Facility Replenishment Cycle*, FCCC/SBI/2007/21, para. 75.
28 International Energy Agency (2009), *World Energy Outlook 2009*, p. 44.
29 Ibid., p. 200.
30 Ibid., p. 281. Later IEA reports argue for even higher dollar amounts.

effort by the FCCC parties to incorporate a systematic analysis of global financial need into the mainstream negotiations on mitigation.

In the lead-up to the Cancun conference in 2010, the UN Secretary-General appointed a High-Level Advisory Group on Climate Change Financing to study potential sources of revenue to meet the commitments on finance. The Advisory Group concluded that the largest portion of revenue would have to come from pricing emissions globally:

> Based on a carbon price of US$20–US$25 per ton of CO_2 equivalent, auctions of emission allowances and domestic carbon taxes in developed countries with up to 10 per cent of total revenues allocated for international climate action could potentially mobilize around US$30 billion annually. Without underestimating the difficulties to be resolved, particularly in terms of national sovereignty and incidence on developing countries, approximately US$10 billion annually could be raised from carbon pricing international transportation, assuming no net incidence on developing countries and earmarking between 25 and 50 per cent of total revenues. Up to US$10 billion could be mobilized from other instruments, such as the redeployment of fossil fuel subsidies in developed countries or some form of financial transaction tax, though diverging views will make it difficult to implement this universally.[31]

In the best case, these methods would raise half the required annual amount for mitigation and adaptation—around $50 billion per year. The case assumes a price for a tonne of carbon dioxide much higher than what it was at the time of writing (when it was about a dollar per tonne) and that cap-and-trade or equivalent systems for raising revenue would be established in all major industrialized economies (when the opposite has happened in Australia, for example).

The FCCC conference at the end of 2011 in Durban did not make progress in clarifying the longer-term sources of funding. It set up a work programme on long-term finance to 'analyze options for the mobilization of resources from a wide variety of sources' and guide Annex I parties in their efforts to mobilize climate finance through to 2020.[32] In the following year, the parties decided to extend the work programme for another year.[33] The Warsaw conference in 2013 set out milestones on long-term climate finance for the period 2014–2020. They include biennial submissions by Annex I parties reporting on their progress in scaling up climate finance through to 2020;[34] and a biennial 'ministerial dialogue'

---

31 Meles Zenawi and Jens Stoltenberg (5 November 2010), *Report of the Secretary-General's High-Level Advisory Group on Climate Change Financing*, pp. 5–6.
32 FCCC (2011), *Decision 2/CP.17*, paras 127, 130.
33 FCCC (2012), *Decision 4/CP.18, Work Programme on Long-Term Finance*, FCCC/CP/2012/8/Add.1, para. 2.
34 FCCC (2013), *Decision 3/CP.19, Long-Term Climate Finance*, FCCC/CP/2013/10/Add.1, para. 10. The parties also adopted a common tabular format for Annex I biennial reporting on climate finance. See Smita Nakhooda et al. (2013), *Mobilising International Climate Finance: Lessons from the Fast-Start Finance Period*, p. 8.

on climate finance starting in 2014 and ending in 2020. Since the Copenhagen Accord did not specify any amounts to be raised in the years 2014–2019 (it only mentioned the amount for 2020), the FCCC conference sought to fill the lacuna with the imprecise instruction that Annex I parties are 'to maintain continuity of mobilization of public climate finance at increasing levels from the fast-start finance period', i.e. from the $30 billion to the $100 billion.[35]

In the same year that finance was promised, the FCCC parties established the Green Climate Fund as a new operating entity of the financial mechanism of the Convention.[36] The GCF is fully under the FCCC's control and is responsible for both adaptation and mitigation funding. It is to promote 'the paradigm shift towards low-emission and climate-resilient development pathways',[37] while being accountable to, and subject to the guidance of, the Conference of the Parties.[38] It is governed by a 24-member board comprising an equal number of Annex I and non-Annex I members.[39] A share of the promised funding through to 2020 and beyond is to flow through the GCF.[40] The relationship between GCF and the GEF has not yet been clarified.[41]

The GCF will have to engage in fund-raising if it is to net some of the promised international finance: 'The COP will make assessments of the amount of funds that are necessary to assist developing countries in implementing the Convention, in order to help inform resource mobilization by the GCF'.[42] The GCF Board will start with 'an initial resource mobilization process' and later transition to formal replenishment rounds.[43] A steady flow of funds to the GCF is thus not guaranteed, and practically speaking could not be agreed on, prior to an agreement on burden-sharing. It will no doubt take a while before the GCF builds any financial clout. By June 2013, states had pledged a total of $9 million (which of course is an insignificant amount) to the GCF Trust Fund.[44] At the Warsaw conference in December 2013, the plenary of the FCCC parties called for 'ambitious and timely contributions by developed countries to enable an effective operationalization [of the GCF] that reflects the needs and challenges of developing countries in addressing climate change'.[45] Acknowledging that the GCF was still a fund in

---

35 FCCC (2013), *Decision 3/CP.19*, para. 7.
36 FCCC (2011), *Decision 3/CP.17*, para. 3.
37 Ibid., Annex, para. 2.
38 Ibid., Annex, para. 4.
39 Ibid., Annex, para. 5. The World Bank serves as the trustee of the fund: ibid., Annex, para. 26.
40 FCCC (2010), *Decision 1/CP.16*, para. 100.
41 See Yulia Yamineva and Kati Kulovesi (2013), 'Framework for Climate Finance', p. 217.
42 FCCC (2013), *Decision 5/CP.19, Arrangements between the Conference of the Parties and the Green Climate Fund*, FCCC/CP/2013/10/Add.1, Annex, para. 17.
43 Green Climate Fund (7 November 2013), *Report of the Green Climate Fund to the Conference of the Parties*, FCCC/CP/2013/6, para. 29.
44 Ibid., para. 42; after disbursements, the balance stood at around $2 million: World Bank (2012–2013), *The World Bank Group Modified Cash Basis Trust Funds*, p. 70.
45 FCCC (2013), *Decision 4/CP.19, Report of the Green Climate Fund to the Conference of the Parties and Guidance to the Green Climate Fund*, FCCC/CP/2013/10/Add.1,

name only, the plenary 'Underline[d] that initial resource mobilization should reach a very significant scale' as a matter of urgency.[46]

So far, I have reviewed the main FCCC provisions laying down the obligations of Annex I parties in relation to financial support for mitigation. While there has been no agreed basis for determining how much support should be provided, as of 2010 Annex I parties had become committed to the mobilization of round-figure, but nevertheless quantified, amounts up to 2020. The only legal impetus for state action was the FCCC's general mitigation rule. States were hesitant to elaborate it in accordance with its inherent logic. They were instead devising ad hoc responses to it, in relation to both mitigation itself and, as we have seen in this chapter, finance for mitigation. However imprecisely, these two elements had become quantified for the 2010–2020 period, and this was an advance. Yet the pledges on mitigation and finance for mitigation were only vaguely linked to each other through the NAMA process and the GCF, and they were routinely treated as separate negotiation items, even though both have one and the same objective, namely mitigation.[47] Global emission and finance budgets, as well as proposals on how to burden-share such budgets, were hardly being discussed at all.

According to article 11.5 of the FCCC (and article 11.3 of the Protocol), Annex I parties may transfer financial resources to developing countries for treaty implementation through bilateral, regional, or multilateral channels.[48] In other words, they need not go through the GEF or the GCF. This makes financial support difficult to aggregate and compare. Another difficulty arises from the requirement of FCCC article 4.3 that climate finance must be 'new and additional'. In all national reporting rounds, the same kind of complaint is constantly raised by those reviewing the information submitted by states: 'only a few Parties explained how they had determined which resources were new and additional';[49] or 'a number of Parties provided information on new and additional financial resources, but the

---

para. 13.
46 Ibid., para. 14.
47 The link between mitigation and finance for mitigation was being resisted in the ADP negotiations on a post-2020 agreement. At the March 2014 ADP meetings, the BASIC group maintained that NDCs by Annex I parties should include information on financial support: International Institute for Sustainable Development (2014), 'Bonn Climate Change Conference: 10–14 March 2014', p. 4. China went further to propose that the post-2020 agreement should have a schedule listing specific amounts to be provided by Annex I parties over a specified timeframe to the GCF (ibid., p. 8). The United States and other Annex I parties opposed these ideas (ibid.). Yet it is a logical consequence of the general mitigation rule that mitigation actions and *finance* for mitigation actions should be treated jointly, because their outcomes are the same.
48 'Multilateral' financing channels include multilateral development banks, such as the World Bank; UN agencies, such as the UNDP and UNEP; and special international agencies such as the GEF. A 'bilateral' financing arrangement is set up by a national government for the purpose of giving aid or investing in a developing country.
49 FCCC Secretariat (1 October 1998), *Second Compilation and Synthesis of Second National Communications from Annex I Parties: Summary*, FCCC/CP/1998/11, paras 44–45.

criteria for determining these resources differed'.[50] A third difficulty is that article 4.3's reference to 'full incremental costs' has not yet been defined by the FCCC parties.[51]

The parties to the FCCC agree that mitigation finance should be subject to transparent reporting, in the same way that greenhouse gas emissions are reported on under the rule on accountable reporting. Yet they have not developed a verification system for mitigation finance (or financial support in general) to match the system for the reporting of emissions. '[T]here is currently no clarity on which body would undertake such verification function and how this would be linked, if at all, to compliance.'[52] Expert Review Teams check that the transfer of finance is reported on by Annex I parties in their national communications,[53] but no equivalent to the IPCC methodologies on emission reporting has been developed for ERTs to apply to the reporting on finance. Verification of financial-transfer obligations has remained superficial.[54] Thus, both in substance (how much money must be raised and who will raise what amount) and in process (how parties will report on finance and how claims will be checked) mitigation finance law is deficient in design, which naturally leads to deficiencies in implementation.

As for the amounts raised in recent years, the GEF claims that by 2013 it had funded 888 projects under its Climate Change Focal Area[55] for a total cost of $3.8 billion.[56] A review was carried out of 113 of these projects. The review found that for 77 per cent of the projects there was evidence of a reduction in emissions.[57] The GEF estimated the total amount of mitigation from the Focal Area projects to be 10.8 Gt CO_2 eq. (consisting of 2.6 Gt CO_2 eq. in direct reduction and 8.2 Gt CO_2 eq. in indirect reduction).[58] The current GEF funding cycle (the Fifth Replenishment, 2010–2014) has an overall pledged amount from 34 countries of $3.54 billion. Of this, the Climate Change Focal Area is due to receive $1.14 billion.[59]

Olbrisch and colleagues refer to estimates that the total amounts of different types of international funding provided to non-Annex I parties for mitigation and adaptation in the 2006–2009 period was $10–15 billion per year, more than three-

---

50 FCCC Subsidiary Body for Implementation (16 May 2003), *Compilation and Synthesis of Third National Communications from Annex I Parties: Executive Summary*, FCCC/SBI/2003/7, para. 56.
51 See Susanne Olbrisch et al. (2011), 'Estimates of Incremental Investment', p. 971.
52 Yulia Yamineva and Kati Kulovesi (2013), 'Framework for Climate Finance', pp. 214–215.
53 See Chapter 2.
54 Taryn Fransen (2009), *Enhancing Today's MRV Framework to Meet Tomorrow's Needs: The Role of National Communications and Inventories*, p. 7 ('Annex I countries typically report extensive [financial-transfer] data ... However, the utility and comparability of this information is limited').
55 Global Environment Facility (2013), *Final Report of the Fifth Overall Performance Study of the GEF: At Crossroads for Higher Impact*, GEF/R.6/17, p. 6.
56 Ibid., p. 8.
57 Ibid., p. 70.
58 Ibid., pp. 50–51.
59 Global Environment Facility, <www.thegef.org>.

quarters of which was for mitigation.⁶⁰ Yamineva and Kulovesi write that bilateral assistance for climate change purposes is estimated at $5.8 billion per year, while non-FCCC climate funds (from UN agencies and other international bodies) provided around $2 billion annually. Other sources of multilateral assistance (such as the Climate Investment Funds, established in 2008) provided around $3 billion per annum. Private-sector investment, most of which goes to renewable energy, is much higher at around $35 billion a year.⁶¹ The usefulness of such information is questionable, but it gives an indication of the level of financial transfer for mitigation at the turn of the decade. The amounts were miniscule compared with the projected needs in 2020, but also compared with the amount of mitigation finance that stays within the borders of the Annex I group and does not cross to developing countries. The IPCC refers to assessments of all current annual financial flows whose expected effect is to reduce emissions or enhance adaptation to climate change and variability as amounting to $343 to $385 billion per year globally.⁶² Out of this, the total public climate finance that flowed to developing countries was between $35 and $49 billion per annum in 2011 and 2012. Estimates of private climate finance crossing to developing countries over the period 2008–2011 range from $10 to $72 billion per annum, although they include foreign direct investment as equity and loans in the range of $10 to $37 billion a year.⁶³

Most of the information above relates to the fast-start finance period under the FCCC, namely 2010–2012, during which time $30 billion was to be raised by Annex I parties, half of it for mitigation. With quantified promises made for the first time, observers began to pay closer attention to matters under article 4.3 of the FCCC. Submissions in 2011 from ten Annex I parties that had budgeted their pledges for the fast-start finance period suggested that most of the funding would come from public sources.⁶⁴ By the end of the period, the United States claimed to have provided $7.5 billion of the promised amount, consisting of around $4.7 billion of Congressionally appropriated assistance and $2.7 billion in the form of US development finance or made available through export-credit agencies.⁶⁵ To take another example, Australia's fast-start finance commitment of around $600 million would be 'Drawn from a growing aid budget [and] it

---

60 Susanne Olbrisch et al. (2011), 'Estimates of Incremental Investment', p. 979.
61 Yulia Yamineva and Kati Kulovesi (2013), 'Framework for Climate Finance', p. 192.
62 IPCC (2014), *Climate Change 2014: Mitigation of Climate Change: Working Group III Contribution to the Fifth Assessment Report of the Intergovernmental Panel on Climate Change: Summary for Policymakers*, p. 27.
63 Ibid., p. 27.
64 Most submissions claimed that the resources were either new and additional or they helped to mobilize new and additional funds from other sources. FCCC Secretariat (15 August 2011), *Submissions on Information from Developed Country Parties on the Resources Provided to Fulfil the Commitment Referred to in Decision 1/CP.16, Paragraph 95*, FCCC/CP/2011/INF.1.
65 US Department of State (2012), *Meeting the Fast Start Commitment: US Climate Finance in Fiscal Year 2012*, p. 1.

does not displace funding from existing aid programs'.[66] This is also an example of how superficially states approach the test in article 4.3 of the FCCC of 'new and additional'.[67] Australia said nothing more about the test than what has been quoted. The country's aid budget has shrunk each year since 2012,[68] underscoring the unreliability of transfer commitments.

Nakhooda and colleagues have examined the claims made for the fast-start finance period. They found that Annex I parties reported that they had raised $35 billion for developing countries in the relevant period, exceeding the target by $5 billion.[69] Germany, Japan, Norway, the United Kingdom, and the United States accounted for nearly 80 per cent of the total contribution.[70] However, the study found that not all of the funding was new or additional. Several baselines could have been used by states to assess whether the fast-start finance commitments were 'new and additional', but because there has been no agreed baseline, each country adopted its own criteria. Many states reported projects, programmes, and funds as fast-start finance that were already being supported by them before 2010.[71] About half of the $35 billion comprised loans, guarantees, and insurance, including export-credit finance for companies based in Annex I parties to invest in developing countries. The Nakhooda study notes that this kind of development finance is quite different from traditional development assistance which is in the form of grants or concessional loans that have an immediate and direct cost to donor budgets. Guarantees, loans, and insurance do not.[72] Nevertheless, the study concludes that a significant proportion of fast-start finance probably reflects real increases in support over the preceding period.[73]

A substantial amount of the fast-start finance was channelled through intermediaries, including dedicated funds and multilateral development banks, as well as through bilateral arrangements (about two-thirds of the total) and state development agencies.[74] Nakhooda and colleagues remark on the lax reporting practices for fast-start finance and contrast them with the stringent reporting regime for greenhouse gas emissions and mitigation targets. They argue that reporting on financial transfers needs to be improved through the adoption of disaggregated reporting, increased labelling of the funds supplied, and the introduction of clarity about objectives, channels, and instruments. Until such reforms are implemented,

---

66 Australian Government (2011), *Australia's Fast-Start Finance: Progress Report*, p. 1.
67 Taryn Fransen (2009), *Role of National Communications and Inventories*, p. 7 ('major Annex I parties ... routinely fail to report on how this determination [of new and additional] has been made. During the [Expert Review Team] review process, the United States indicated that it considered all funding in any year to be "new and additional"').
68 See e.g. Karen Barlow, 'Budget 2014: Axe Falls on Foreign Aid Spending, Nearly $8 Billion in Cuts Over Next Five Years', <http://www.abc.net.au/news/2014-05-13/budget-2014-axe-falls-on-foreign-aid-spending/5450844>.
69 Smita Nakhooda et al. (2013), *Lessons from the Fast-Start Finance Period*, p. i.
70 Ibid., p. 39.
71 Ibid., p. ii.
72 Ibid., pp. 39–40.
73 Ibid., p. 26.
74 Ibid., p. iii.

doubt will remain about whether Annex I parties are meeting their commitments on mitigation finance.[75] Other scholars have echoed this criticism, with Yamineva and Kulovesi calling the information from FCCC parties on fast-start finance 'highly approximate' because of the reporting deficiencies.[76]

The problem is not confined to climate finance. A UN 2011 report underscored that no reliable international environmental financial tracking system exists, making it difficult to assess the total amount invested in environmental activities through global transfers, even where the scope of the inquiry is limited to the UN system.[77] Understanding the global transfer of climate finance is clearly more important than understanding the transfer of finance for environmental activities in general, because mitigation finance is tied up with the effort to avoid 2°C of warming. In 2013, the FCCC's Work Programme on Long-Term Finance reported on meetings it had held with state parties and other stakeholders. The report noted that two of the obstacles for the mobilization of climate finance by Annex I parties were a lack of agreement on common definitions of what comprises climate finance and a lack of agreement on how to track and report private-sector finance, specifically its attribution to a particular country.[78] The report also acknowledged that no agreed frameworks had been developed for reporting on fast-start finance, and consequently parties resorted to using different methodologies. Disputes had arisen when several developing countries alleged apparent discrepancies between the amounts of fast-start finance reported to have been provided and the amounts received. The report called for improved transparency in reporting and suggested that the biennial reports, which will account for both mitigation actions and climate finance (as well as undergo IAR/ICA), may help address the problem.[79]

The United States, in its submissions on the post-2020 agreement, called for greater transparency on financial transfer, as 'it was not always clear how much finance was flowing from whom to whom'. Yet it also supported 'Significantly more attention ... to private sources of funding, including ways in which public resources and policies can help mobilize such funding',[80] even though this would mean less transparency. The US submissions recalled and defended the Copenhagen Accord's language, emphasizing that 'it calls for mobilization, rather than provision, of funds'. The United States concluded that 'These institutional

---

75 Ibid., p. 43.
76 Yulia Yamineva and Kati Kulovesi (2013), 'Framework for Climate Finance', p. 211. See also Susanne Olbrisch et al. (2011), 'Estimates of Incremental Investment', p. 979; Haroldo Machado-Filho (2012), 'Financial Mechanisms under the Climate Regime', in *Promoting Compliance in an Evolving Climate Regime*, ed. Jutta Brunnée, Meinhard Doelle, and Lavanya Rajamani, p. 234; and Catherine Redgwell (2012), 'Facilitation of Compliance', in *Promoting Compliance in an Evolving Climate Regime*, ed. Jutta Brunnée, Meinhard Doelle, and Lavanya Rajamani, p. 185.
77 Governing Council of the United Nations Environment Programme (2011), *Environment in the United Nations System*, UNEP/GC.26/INF/23.
78 Work Programme on Long-Term Finance (1 November 2013), *Report on the Outcomes of the Extended Work Programme on Long-Term Finance*, FCCC/CP/2013/7, Annex, para. 14.
79 Ibid., Annex, para. 21.
80 United States (2014), *US Submission on Elements of the 2015 Agreement*, p. 9.

and other advances in climate finance will continue to be relevant in the post-2020 period.'[81] Yamineva and Kulovesi note that most developing countries 'insist that the [promised Copenhagen] funding should mainly come from public sources in developed countries and be channeled through the FCCC. ... Developed countries [by contrast] are keen to avoid strong and prescriptive language on public funding'.[82]

Thus while improved reporting on climate finance is important for compliance with the general mitigation rule, on the substantive side questions about the sources of long-term finance, how the sufficiency of the amounts will be defined, and how responsibility will be allocated to individual Annex I parties are matters that remain poorly developed. The co-chairs of the Work Programme on Long-Term Finance correctly observed that the overriding priority is that 'sufficient finance becomes available as soon as possible to support mitigation ... actions consistent with keeping the global average temperature below 2°C'.[83] However, they continued, neither is the $100 billion goal clearly defined ('the sources and flows of climate finance that can be regarded as contributing to meeting this goal have not been agreed on') nor has there been any 'agreement on burden-sharing' among Annex I parties on how to achieve the funding target.[84] The co-chairs remarked that the issue of burden-sharing among Annex I parties is a political issue that could not be resolved within the Work Programme.[85] Yet there is no other FCCC forum for it to be resolved within.

5.3 Conclusion on the financial support rule

Because mitigation costs are lower in developing countries, the extent of financial transfer to support mitigation policies and projects in developing countries could be a measure of the extent to which developed countries are serious about seeking emission reductions in general. At the same time, problems with transparency and accountability in both the provision and the use of mitigation funding are barriers to understanding how the FCCC parties are meeting their funding obligations. Moreover, funding promises have been paltry, ad hoc, and largely unsourced. The evidence reviewed here suggests that the FCCC parties have failed to comply with the general mitigation rule to the extent that it subsumes the Annex I states' obligations on mitigation finance.

---

81 Ibid., p. 10.
82 Yulia Yamineva and Kati Kulovesi (2013), 'Framework for Climate Finance', p. 209.
83 Work Programme on Long-Term Finance (1 November 2013), *Outcomes of the Extended Work Programme*, Annex, para. 19.
84 Ibid., Annex, para. 24.
85 Ibid., Annex, para. 15.

6 Climate law and 'optional' mitigation mechanisms

6.1 Introduction: Expanding the investigation of regime rules

The earlier chapters have distilled an international climate law and assessed state performance against it. First among the rules of climate law is accountable reporting, which I have also called the rule on transparency about emissions and mitigation actions. It is a universally recognized if not uniformly practised rule. It has normativity, as it rests on the conviction that the problem of climate change, with its unique features of universal causation and responsibility, and its potential to cause universal catastrophe, cannot be effectively addressed unless each country's contribution to the problem is understood and continually tracked. The rule, moreover, mounts a permanent attack on free riding, which is essential if states are to commit to emission reductions. The second rule, on prevention, is more complex in form and content. This general mitigation rule demands of states that they reduce emissions to a level that avoids dangerous climate change. It has several implications for states, the most important of which is an obligation to agree on a fair individualization of the mitigation burden, referenced to a global emission budget consistent with prevention. My analysis has also suggested that the law on financial transfer is best understood as an aspect of the general mitigation rule, because the imperative behind it is the mitigation imperative itself. Compliance with the prevention rule (in terms of both emission reductions and financial transfer) has been very weak to date.

The purpose of the present chapter is not to expand the rulebook. Climate law hopefully will develop with time, but at the moment there is, alas, nothing more to highlight about it. In this chapter, I will examine how the rules on transparency and prevention apply to specialized, project-focused, mitigation areas of the FCCC and Kyoto Protocol. The question addressed here is about how the two general rules are observed (or not) outside the exact contexts in which they were developed. In the case of the Protocol, I will use the CDM to focus the examination. From the FCCC's side I will consider REDD, a mitigation measure of great potential. The CDM and REDD have similarities (e.g. both rely on the concept of additionality), and differences (not least that the CDM is fully operational whereas REDD is not). States are not obliged to participate in the CDM, and not all states are obliged to participate in REDD, which is why I call them 'optional'.

136 *'Optional' mitigation mechanisms*

When the Kyoto Protocol laid the foundations for its three flexibility mechanisms it was understood that each would proceed to establish a body of rules and administration to manage its own affairs as if they were separate businesses, while remaining under the guidance of the Protocol parties. State participation in the flexibility mechanisms and in emission trading has always been a matter of choice for each state.

In the first part of this chapter, I will focus on the largest and methodologically most challenging of the Protocol's three mechanisms, the CDM. My purpose is to examine state compliance with the obligations created through the mechanism to the extent that they come under the rule on accountable reporting and the general mitigation rule. When states establish a mechanism like the CDM, they accept, at the very least, an obligation to implement it consistently with the climate change regime's overarching legal rules.

6.2 State obligations under the Protocol's Clean Development Mechanism

The CDM is a baseline-and-credit mechanism. Created under article 12 of the Kyoto Protocol, it enables developing countries to host emission-reduction projects that generate emission allowances (CERs). These can be used by an Annex I party to comply with its assigned amount for a commitment period. CERs are treaty-approved offsets whose effect in theory is to neutralize Annex I domestic emissions, tonne for tonne. This enables the country to emit greenhouse gases above the emission limit it has accepted under the Protocol. It is vital, then, that emission reductions at a CDM project are genuinely additional to any reductions that would have been achieved in the project's host state under business-as-usual conditions, i.e. without the assistance of the CDM.

Additionality is a fundamental rule of the CDM. It is inherent to both its logic and integrity. A second treaty-mandated rule of the CDM is possibly more a matter of politics. CDM projects must assist non-Annex I parties to achieve sustainable development. The two objectives give rise to important compliance considerations for participating states.

6.2.1 Establishment and supervision of CDM projects

Much international and domestic regulation was necessary to make the CDM possible.[1] Article 12 of the Protocol creates an Executive Board under the authority

---

[1] The CDM has been described as densely regulated, with a complex set of methodological rules (see Peter Newell and Matthew Paterson (2010), *Climate Capitalism: Global Warming and the Transformation of the Global Economy*, p. 149); it has also been described as a 'highly dynamic body of legal text': Matthias Krey and Heike Santen (2009), 'Trying to Catch up with the Executive Board: Regulatory Decision-Making and Its Impact on CDM Performance', in *Legal Aspects of Carbon Trading*, ed. David Freestone and Charlotte Streck, p. 232.

of the plenary of the parties to supervise the mechanism.[2] The same article provides for private and public entities to be involved in CDM operations.[3] Private entities, especially the Designated Operational Entities, are given a vital role in the operation of the CDM. A host-state government is required to set up a Designated National Authority to formally approve and contribute to the oversight of CDM projects.[4]

Article 12 calls for a dedicated monitoring, reporting, and verification system for the CDM. The plenary is to 'elaborate modalities and procedures with the objective of ensuring transparency, efficiency and accountability through independent auditing and verification of project activities'.[5] I will briefly review how this has worked out. We should note that there is no mention in the Protocol of a compliance system, as such, for the CDM.

The CDM's project-approval procedure, from registration of a project by the CDM Executive Board through to the issuance of CERs to the account of the project owners, may be summarized as follows.[6] A project developer prepares a Project Design Document using an approved CDM methodology.[7] CDM methodologies are developed by private entities for approval by the Executive Board.[8] They include methods for the demonstration of additionality. The host state's DNA must confirm that the project will make a positive contribution to sustainable development.[9] The proposed project must then be validated by a Designated Operational Entity. Validation is intended as an independent assessment of a project's compliance with CDM rules.[10] The DOE functions as an auditor, approved by the Executive Board, but hired by the project developer. It must verify whether the baseline (or reference case) identified in the design document is 'reasonable',[11] and whether the proposed project activity is additional.[12] If a DOE determines that the requirements for a CDM project have been met, it will, on behalf of the project developer, request the Executive Board to register the project. Registration by the Executive Board constitutes the Protocol regime's

---

2 The supervision function is detailed in Kyoto Protocol (2005), *Decision 3/CMP.1, Modalities and Procedures for a Clean Development Mechanism as Defined in Article 12 of the Kyoto Protocol*, FCCC/KP/CMP/2005/8/Add.1, paras 2–4. On the constitution of the ten-member CDM Executive Board, see ibid., paras 7–12.
3 Charlotte Streck and Jolene Lin (2009), 'Mobilising Finance for Climate Change Mitigation: Private Sector Involvement in International Carbon Finance Mechanisms', 10 (1) *Melbourne Journal of International Law* 70, p. 73.
4 Ibid., p. 79.
5 Kyoto Protocol, art. 12.7.
6 See Kyoto Protocol (2005), *Decision 3/CMP.1*; Clean Development Mechanism (2013), *CDM Project Standard (v. 5.0)*, CDM-EB65-A05-STAN; and Clean Development Mechanism (2013), *CDM Project Cycle Procedure (v. 5.0)*, CDM-EB65-A32-PROC.
7 CDM (2013), *Project Standard*, para. 23.
8 Kyoto Protocol (2005), *Decision 3/CMP.1*, paras 5(d) and (j).
9 CDM (2013), *Project Standard*, para. 71.
10 CDM (2013), *CDM Validation and Verification Standard (v. 5.0)*, CDM-EB65-A04-STAN, para. 17.
11 Ibid., para. 88.
12 Ibid., para. 101.

approval of a CDM project; it is a green light to proceed with implementation.[13] Non-CDM finance is normally necessary at this stage to cover the upfront costs of project development and initial implementation.

The project developer is responsible for monitoring emissions from the project (or, where a project produces no emissions, the developer must monitor the project output that is designed to displace emissions elsewhere), in accordance with the monitoring requirements of the applicable methodology.[14] After a period of operation, another DOE is to verify that emission reductions have taken place in the amount claimed in the project developer's monitoring report.[15] The verification procedure is a condition for the issuance of CERs. It is repeated periodically throughout the life of the project. The DOE's verification report is followed by certification, which is the DOE's assurance to the Executive Board that the emission reductions claimed by the project are 'real'.[16] In the normal course of events, the Executive Board will issue CERs on the basis of the DOE's certification. When the CERs are received by the project developer, they may be sold in a compliance market such as the EU ETS.[17] The proceeds will fund the on-going operation of the project or are used to recoup upfront operating costs.

The lifespan, or crediting period, of a CDM project is 7 years, renewable twice (up to 21 years), or is a fixed period of up to 10 years, after which international support for the project ceases. Different project types have different permissible crediting periods.[18]

The procedure I have outlined has been diversified over the years. There is, for example, internal 'differentiation',[19] in accordance with which 'small-scale' projects (about 40 per cent of CDM projects are small-scale) benefit from a simplified establishment procedure, meaning that they can get up and running faster than regular-size projects. The original CDM model created stand-alone projects. Now, the CDM also allows for 'programmes of activities', under which an unlimited number of similar component project activities in a country can be registered under a single umbrella.[20] Certain project types are excluded from the

---

13 CDM (2013), *Project Cycle Procedure*, para. 97.
14 Ibid., paras 177f.
15 CDM (2013), *Validation and Verification Standard*, para. 244. As Lund notes, the same DOE is generally forbidden from performing both the validation and the verification of the same project, in order to reduce the DOE's incentive to validate projects just to get paid for the later verification tasks: Emma Lund (2010), 'Dysfunctional Delegation: Why the Design of the CDM's Supervisory System Is Fundamentally Flawed', 10 *Climate Policy* 277, p. 282.
16 CDM (2013), *Validation and Verification Standard*, para. 286.
17 Voluntary cancellation of CERs is another possibility, which has led to some demand from the private sector. See CDM Executive Board (2013), *Annual Report*, FCCC/KP/CMP/2013/5 (Part I), p. 8.
18 CDM (2013), *Project Standard*, para. 59. Longer crediting periods apply to afforestation and reforestation projects: ibid., para. 128.
19 Stefan Bakker et al. (2011), 'The Future of the CDM: Same Same, but Differentiated?', 11 *Climate Policy* 752, p. 754.
20 CDM Executive Board (2011), *Annual Report*, FCCC/KP/CMP/2011/3 (Part I), p. 4.

CDM (e.g. nuclear power projects)[21] or their emission allowances have a time-limited use (CERs from forestry projects fall into this category).[22] Some early limitations on project types have been eased. For example, carbon capture and storage was recently allowed as a project type, although no such project had been established at the time of writing.[23]

As I noted earlier, the CDM has no separate compliance mechanism over and above the 'verification' elements that form part of the procedure summarized above. The Executive Board is responsible for most compliance matters, many of which it delegates.[24] In particular, a large part of the responsibility for the mechanism's integrity is delegated to DOEs.[25] The Executive Board tends to follow a DOE's advice in matters of project registration and CER issuance.[26] The Board also has a responsibility to ensure that the rule on additionality is complied with—but, as I will explain below, not the rule on sustainable development. The latter is one area of state obligation that has been neglected in the CDM's design.

---

21 Nuclear-power generation projects have been excluded from the CDM because of safety, security, and political sensitivities. Half-hearted negotiations to lift the CDM's restriction on nuclear-energy projects are continuing; see International Institute for Sustainable Development (2011), 'SB 34 and AWG Highlights: Tuesday, 14 June 2011', 12 (510) *Earth Negotiations Bulletin* 1, p. 2.

22 Forestry-based CDM projects have been of narrow scope. The CDM Executive Board issues temporary rather than permanent CERs for forestry activities (from afforestation and reforestation projects only) as a precaution for dealing with forest impermanence (the euphemism used is 'reversal of storage'). Such offsets have special names under the CDM—*temporary CERs* (tCERs) and *long-term CERs* (lCERs)—to distinguish them from normal CERs. The temporary nature of these CERs means that the Annex I buyer has to repurchase them or substitute them with permanent CERs at the end of a commitment period. See Kyoto Protocol (2005), *Decision 5/CMP.1, Modalities and Procedures for Afforestation and Reforestation Project Activities under the Clean Development Mechanism in the First Commitment Period of the Kyoto Protocol*, FCCC/KP/CMP/2005/8/Add.1. See also Oscar van Vliet, André Faaij, and Carel Dieperink (2003), 'Forestry Projects under the Clean Development Mechanism?', 61 *Climatic Change* 123; and Sebastian Scholz and Ian Noble (2005), 'Generation of Sequestration Credits under the CDM', in *Legal Aspects of Implementing the Kyoto Protocol Mechanisms: Making Kyoto Work*, ed. David Freestone and Charlotte Streck. For the Kyoto Protocol's second commitment period, the total emissions above a party's assigned amount offset by CERs produced from CDM afforestation and reforestation activities is not to exceed *one per cent* of the base-year emissions of that party, multiplied by the duration of the commitment period in years: Kyoto Protocol (2011), *Decision 2/CMP.7, Land Use, Land-Use Change and Forestry*, FCCC/KP/CMP/2011/10/Add.1, para. 19.

23 CDM Executive Board (2013), *Annual Report*, p. 14.

24 On the role of the FCCC Secretariat in this context, see Maria Netto and Kai-Uwe Barani Schmidt (2005), 'CDM Project Cycle and the Role of the UNFCCC Secretariat', in *Legal Aspects of Implementing the Kyoto Protocol Mechanisms: Making Kyoto Work*, ed. David Freestone and Charlotte Streck.

25 Charlotte Streck and Jolene Lin (2009), 'Private Sector Involvement in International Carbon Finance', p. 83 ('Thus, much of the power to create CERs rests with the DOEs').

26 Kyoto Protocol (2005), *Decision 3/CMP.1*, Annex, paras 64–65.

From the climate regime's perspective, the CDM is a mitigation mechanism in two senses; it shifts emission reductions to locations where their cost per unit is lower, and it facilitates more ambitious mitigation pledges by Annex I parties. From a commercial perspective, the CDM is an opportunity for profit. A CDM project developer has an interest in arguing for a fixed long-term baseline trajectory, because this would enable the project to produce 'additional' emission reductions for many years. The host state benefits from the increased economic activity brought to the country by the project, and thus has an interest in the project going ahead. Most buyers of CERs are industries in the EU ETS, so it is a safe assumption that they do not have an interest in the integrity of the emission reductions, but only in the property right to the CERs. Above all, buyers are concerned about price, and the greater the supply of CERs, the lower the price.

Project developers select and pay the DOEs that assess the additionality of their projects and the amount of emission reductions resulting from them. DOEs have an economic incentive to let projects get through, as Lund has noted.[27] The Executive Board can revoke a DOE's delegation by suspending or withdrawing its accreditation.[28] If a review reveals that the Board issued excess CERs to a project on the basis of a DOE's certification, the Board may demand of the DOE that it purchase and transfer to the Board for cancellation an amount of CERs equal to the excess issued.[29] However, 'information asymmetry'—the condition whereby if the Board does not repeat the work of the DOE it cannot be sure whether its verification or validation reports are correct—makes it difficult for the Executive Board to know when to apply its disciplinary power.[30] The Board has arranged for 'spot checks' to get around the asymmetry. In 2009, two such checks were performed; both resulted in the suspension of the DOEs involved for 'systematic non-compliance'.[31] Public documentation from later years does not refer to any further spot checks or suspensions being carried out.

A credible supervisory system is necessary to uphold the CDM's integrity as a mitigation mechanism and to ensure that its implementation does not lead to an increase in global emissions. There has been a considerable effort by NGOs to monitor the CDM and expose its shortcomings.[32] However, the integrity of the process must finally be the responsibility of the Kyoto Protocol parties. They have, in their majority, opted to participate in the CDM,[33] and as a result have become bound by its two central rules (additionality and sustainable development).

---

27 Emma Lund (2010), 'Dysfunctional Delegation', pp. 277, 281.
28 Kyoto Protocol (2005), *Decision 3/CMP.1*, Annex, para. 21.
29 Ibid., para. 22.
30 Emma Lund (2010), 'Dysfunctional Delegation', p. 281.
31 CDM Executive Board (2009), *Annual Report*, FCCC/KP/CMP/2009/16, paras 37–38.
32 See especially Carbon Market Watch, <http://carbonmarketwatch.org/issues-in-the-cdm/>; and Steinar Andresen and Lars H. Gulbrandsen (2005), 'The Role of Green NGOs in Promoting Climate Compliance', in *Implementing the Climate Change Regime: International Compliance*, ed. Jon Hovi, p. 179.
33 As we saw in Chapter 4, most Annex I parties to the Protocol will not need to use CERs to stay within their assigned amount for the first commitment period. However,

The CDM was launched in 2005.[34] During the mechanism's first few years, project-creation was modest. Since 2010 the approval rate of projects has accelerated. At the time of writing, over 7,500 projects had been established in at least 90 countries. The total investment in registered or soon-to-be-registered CDM projects, as of June 2012, was estimated at $215.4 billion.[35] By mid-2014, the CDM had issued 1.46 billion CERs.[36] The revenue generated from the sale of CERs exceeds $10 billion.[37] The Executive Board has estimated that it will issue a further 6 billion CERs by 2020.[38] Wealthy industrialized countries—mainly those in the European Union, in whose carbon market most CERs are sold—are estimated by the Board to have saved more than US$3.6 billion in compliance costs due to the CDM.[39]

6.2.2 Additionality rule

A CDM project must act as a sink for, or destroy, greenhouse gases, or create a product or service that substitutes itself for (i.e. displaces) an existing or planned and comparatively more emission-intensive product or service. These matters are verifiable. The additionality element is epistemologically more complex. The Protocol requires CDM projects to deliver mitigation benefits that are 'additional to any that would occur in the absence of the certified project activity'—and that are measurable.[40] A decision of the Protocol parties defines additionality as follows: 'A CDM project activity is additional if anthropogenic emissions of greenhouse gases by sources are reduced below those that would have occurred in the absence of the registered CDM project activity.'[41] That which is to be 'measurable' is the difference between an actual and hypothetical emission trajectory. Since there are no quantified emission limits for developing countries, a CDM host state's future emissions cannot be known in advance but must be hypothesized.

To meet the additionality requirement, a project developer must demonstrate that the proposed project faces 'implementation barriers'.[42] The first consideration here is whether there exist alternatives to the proposed project that do not reduce emissions, and that could go ahead unimpeded by the applicable laws and regulations of the host state. The second question about barriers considers whether an 'investment analysis' establishes that the proposed project either

---

CERs may prove more relevant to compliance in the second commitment period.
34 A handful of projects were registered as early as 2004.
35 CDM Executive Board (2012), *Benefits of the Clean Development Mechanism 2012*, p. 8.
36 A further 141,000 CERs (a small number by comparison) had been issued to Programmes of Activities.
37 CDM Executive Board (2012), *Benefits of the CDM 2012*, p. 9.
38 CDM Executive Board (2013), *Annual Report*, p. 7.
39 Ibid., p. 7.
40 Kyoto Protocol, art. 12.5.
41 Kyoto Protocol (2005), *Decision 3/CMP.1*, para. 43.
42 Clean Development Mechanism (2012), *Tool for the Demonstration and Assessment of Additionality (v. 7.0)*, CDM-EB70 Annex 8, para. 2.

is not the most economically or financially attractive or is not economically or financially feasible, compared with the (emission-intensive) alternatives. The third step considers what other barriers exist that prevent the implementation of the proposed project but do not prevent the implementation of at least one of the alternatives. The final question is whether the proposed project faces a 'common practice' barrier, either because it is the first of its kind or because this kind of project is not diffused in the relevant geographical area and therefore counts as unusual practice in the context of that area.[43]

The demonstration of barriers is for the purpose of showing that additional costs are involved in reducing emissions through the implementation of the proposed project, which in turn shows that a subsidy is needed to make up the difference. It must be shown that only the CDM's subsidy is available. This in itself is proof of another barrier: that there is no reasonable source of finance for the proposed project other than the CDM. Without the CDM's support, the 'additionality' reasoning goes, the project could not make ends meet, would not go ahead, no project like it could go ahead with funding from a non-CDM source, and the higher-emitting status quo, or a business-as-usual increase in emissions, would continue. On the other hand, with the CDM's financial support for the proposed project, the emission trajectory of the host state would shift to a lower curve. The difference would represent the emission reductions created by the project. It is then only a matter of converting this difference into allowances for sale to firms or governments that are required to pay for their emissions or cancel excess emissions, thereby generating a subsidy for the project.

Where a project is not of the stand-alone type but is a programme of activities, what needs to be established to prove additionality is that, in the absence of the programme of activities, none of the component project activities would be realized.[44] The difference is one of complexity, but the logic is the same.

A critical element in all of this is the shape of the trajectory for the reference case or baseline.[45] If it is set too high or kept in place for too long, the emission reductions from the CDM project will not be 'real', as required by the Protocol.[46] It is important to notice that the barriers on which the case for additionality is constructed are diachronic. They are presumed to exist not only now, but also in a year from now—and for a 10-year project, in 10 years from the date of certification.

To illustrate some of the problems involved, let us consider a CDM landfill-gas project that captures methane. The reference case emissions would be all the methane that would escape from the landfill without the landfill-gas project

---

43 Ibid., para. 60.
44 Clean Development Mechanism (2013), *CDM Demonstration of Additionality, Development of Eligibility Criteria and Application of Multiple Methodologies for Programmes of Activities (v. 3.0)*, CDM-EB65-A03-STAN, para. 7.
45 See Axel Michaelowa (2005), 'Determination of Baselines and Additionality for the CDM: A Crucial Element of Credibility of the Climate Regime', in *Climate Change and Carbon Markets: A Handbook of Emission Reduction Mechanisms*, ed. Farhana Yamin.
46 Kyoto Protocol, art. 12.5.

operating to capture and burn the gas. The project emissions consist of an allowance for methane leakage and the carbon dioxide that results from burning the captured methane. The difference, projected over the life of the project (several years), is converted into the CERs earned by the project. The difficulty is not the emission reductions that the project may claim from day to day, which could be quite precisely measured; it is the assumption that the methane would not eventually be captured through another initiative made possible by a subsidy or other injection of finance from a source other than the CDM. It may be true that no alternative exists when the project is proposed, however the circumstances a year or two later can only be guessed at. The facts concerning additionality in the present must be extended to future situations about which a lot less is known and a lot more must be assumed. Moreover, the very availability of CDM funding could create a perverse incentive to understate or even conceal the availability of alternative sources of funding. Consider, for example, the possible case of a palm-oil manufacturer that does not currently collect methane from its evaporation ponds because it is not common practice to do so. In proposing a CDM project, this manufacturer must make a case that, 10 years down the track, covering the ponds will still not be common practice—even though, in the absence of the CDM, the manufacturer might have covered the ponds within that decade at its own expense for good publicity.

Additionality is no longer required to be demonstrated in each individual case. The Executive Board maintains a 'positive list' of small-scale project types that qualify for 'automatic additionality' due to their 'obvious ability' to reduce emissions and the 'obvious barriers' they face in implementation.[47] The Board has also developed guidelines on 'standardized emissions baselines' that allow countries to calculate 'typical emissions' for an entire sector and create a list of technologies or measures that are, again, 'automatically additional'. In these and other ways, the Executive Board has sought to make the demonstration of additionality progressively more 'straightforward' for developers.[48]

6.2.3 Sustainable development rule

The Kyoto Protocol states that 'The purpose of the clean development mechanism shall be to assist Parties not included in Annex I in achieving sustainable development'.[49] This has generally been interpreted as a requirement that attaches to each CDM project (i.e. that the project itself must create sustainable development benefits in the host state), as well as to the mechanism as a whole. The requirement is a rule of treaty law like the additionality requirement, however its rationale is not as readily apparent. A likely explanation is political: by making sustainable development a necessary condition of the operation of the CDM, the Protocol parties can claim that the scheme is delivering long-lasting, transformative assistance to developing countries and is not simply serving to

---

47 CDM Executive Board (2013), *Annual Report*, p. 15.
48 CDM Executive Board (2011), *Annual Report*, p. 6.
49 Kyoto Protocol, art. 12.2.

offsett the excess emissions of wealthy countries. Another possible explanation is the UN's general philosophy that financial support for development should be directed at outcomes that improve sustainability.[50]

The standard analysis of 'sustainable development' entails dissecting the concept into three 'dimensions': economic development, social development, and environmental protection.[51] The breakdown has some utility as it suggests that any improvement along one of the dimensions would be evidence of a contribution to sustainable development, as long as the other dimensions do not suffer any detriment. There is no standard definition of sustainable development beyond its three dimensions, and none is provided in the Kyoto Protocol. Several attempts to particularize the definition have been made, e.g. by the CDM Executive Board and scholars (see below).

The responsibility to determine whether a CDM project will contribute to sustainable development at the state level rests with the corresponding host country's Designated National Authority.[52] As described earlier, the DNA attests in a letter of approval whether and how, in its judgement, the proposed CDM project will contribute to the country's sustainable development.[53] After this point in the procedure, it is not clear who retains oversight of the question. The Executive Board periodically reports on the sustainable development benefits of CDM projects,[54] and has provided CDM project developers with advice on how to identify and expand upon such benefits.[55] The Board has nevertheless avoided claiming any oversight responsibility for sustainable development. DOEs—and by extension the Executive Board, given that it has no other way of supervising CDM projects—are explicitly steered away by CDM rules from validating sustainable development benefits claimed by CDM projects.[56] A DNA is said to be the best judge of country-level sustainable development priorities, yet DNAs have not claimed any responsibility for assessing, nor are they known to follow up on, the impacts of the projects they approve.

Decisions of the Protocol parties have failed to elaborate the obligation to ensure that CDM projects enhance sustainable development in host states. No reporting procedure on this matter has been developed. The obligation to ensure

---

50 See the UN's Millennium Development Goals at <http://www.un.org/millenniumgoals/>.
51 Johannes Alexeew et al. (2010), 'An Analysis of the Relationship between the Additionality of CDM Projects and Their Contribution to Sustainable Development', 10 *International Environmental Agreements* 233, p. 236.
52 Kyoto Protocol (2005), *Decision 3/CMP.1*, para. 29.
53 See Section 6.2.1. At the Durban CMP in December 2011, the prerogative of the host states to define their sustainable development criteria in relation to the CDM was reaffirmed: Kyoto Protocol (2011), *Decision 8/CMP.7, Further Guidance Relating to the Clean Development Mechanism*, FCCC/KP/CMP/2011/10/Add.2, para. 5.
54 CDM Executive Board (2011), *Benefits of the Clean Development Mechanism 2011*; and CDM Executive Board (2012), *Benefits of the CDM 2012*.
55 Clean Development Mechanism (2014), *Voluntary Tool for Describing Sustainable Development Co-Benefits of CDM Project Activities or Programmes of Activities (v. 1.1)*, SD-TOOL01.
56 CDM (2013), *Validation and Verification Standard*, para. 50.

that the CDM *programme* is meeting its treaty obligations would seem to rest on the Protocol parties as a whole, although not all Protocol parties participate in the CDM, so this interpretation may lead to difficulty. However, most parties do participate (as host states or purchasers of CERs). Responsibility for compliance with the rule must at least rest with these parties. In sum, the legal situation has been kept obscure.

Rajamani argues that because no international body is tasked with or authorized to inquire into the sustainable development impacts of a CDM project, 'the obligation placed on developing countries to ensure that CDM projects they host contribute to sustainable development does not lend itself to compliance assessment at the international level'.[57] However, the fact that no international body exists with a responsibility to assess this type of compliance does not mean that compliance with the sustainable development requirement of the CDM may be ignored. The lack of a compliance system makes it more difficult to assess compliance with the obligation—but it does not extinguish the obligation.

6.3 Compliance with the CDM's core rules

6.3.1 Additionality made 'simple'

Establishing a workable way by which to judge whether CDM emission reductions are additional to what would have happened otherwise has been described as a 'vexing challenge'.[58] The 'inescapable subjectivity'[59] of the judgement is evident from the CDM's guidelines. A DOE is to determine whether the baseline identified for the proposed project is the scenario that 'reasonably represents' the emissions that would occur in the absence of the project.[60] The test is one of 'reasonableness'. The DOE is to assess whether the list of 'alternatives' presented in the project developer's argument for additionality contains all the plausible alternatives, which, in the DOE's opinion and 'on the basis of its local and sectoral knowledge', present 'a viable means of supplying the comparable outputs or services that are to be supplied by the proposed project'.[61] The test is thus subjective: the subject is the DOE, with its knowledge of the socio-economic circumstances relevant to the proposed project. In assessing whether the barriers alleged by the project developer are real, the DOE is to conduct interviews with relevant individuals (members of industry associations, government officials, local experts, and so on). It is also to check independent sources of data (national legislation, surveys of local conditions, and national or international statistics, etc.) to substantiate the

---

57 Lavanya Rajamani (2012), 'Developing Countries and Compliance in the Climate Regime', in *Promoting Compliance in an Evolving Climate Regime*, ed. Jutta Brunnée, Meinhard Doelle, and Lavanya Rajamani, p. 391.
58 Stephen Meyers (1999), *Additionality of Emissions Reductions from Clean Development Mechanism Projects: Issues and Options for Project-Level Assessment*, Research Paper LBNL-43704, p. iii.
59 Ibid., p. iii.
60 CDM (2013), *Validation and Verification Standard*, para. 88.
61 Ibid., para. 114.

existence of the alleged barriers.[62] Thus some substantiation with some objective evidence is necessary. However,

> Since not all barriers present an insurmountable hurdle to a project activity being implemented, the DOE shall apply its local and sectoral expertise to judge whether a barrier or set of barriers would prevent the implementation of the proposed project activity and would not equally prevent implementation of at least one of the possible alternatives, in particular the identified baseline scenario.[63]

Additionality finally comes down to the judgement of one DOE.

The CDM guidelines require that national and sectoral policies and circumstances be taken into account in the establishment of a baseline scenario, yet 'without creating perverse incentives that may impact host Parties' contributions to the ultimate objective of the Convention'.[64] The guidelines do not explain how such incentives could be avoided. I noted earlier that the baseline is fixed for the full crediting period of the project. It is only when a request is made to renew the crediting period (e.g. by extending a seven-year project by another seven years) that the project owner is required to update the information on the baseline and draw up a new baseline for the next period.[65]

All CDM methodologies make obligatory the use of multipliers that increase the conservativeness of estimates of avoided emissions.[66] Increased conservativeness must be distinguished from conservativeness itself. Increasing conservativeness does not make the final estimation of emission reductions conservative in the sense of being 'real'. It is impossible to say with certainty what would have been the case were it not for the CDM project. The multipliers do no more than reduce the probability of error. The CDM is by its nature *non*-conservative, which is why NGOs have opposed it from the very start. Not all CDM projects face the same level of uncertainty about the baseline. Those that are clearly not financially attractive, and are therefore aimed at 'high-hanging fruit', are very unlikely to have current alternatives—but the unlikelihood decreases with every year of the projection.[67] In rhetoric, the CDM is about additionality, but in practice it is about encouraging the creation of more and more projects to reap the low-hanging fruit without unduly constraining investment or the flow of climate finance. Additionality is treated not as an end in itself but as a factor in a balancing exercise between environmental integrity, on the one hand, and the availability of attractive incentives for mitigation finance, on the other.

---

62 Ibid., para. 126.a.
63 Ibid., para. 126.b.
64 CDM (2013), *Project Standard*, para. 43.
65 Ibid., para. 230.
66 Kyoto Protocol (2005), *Decision 3/CMP.1*, para. 45(b).
67 Stephen Meyers (1999), *Additionality of Clean Development Mechanism Projects*, p. 4.

For these reasons, it can be assumed that many non-additional CDM projects have been established. Carbon Market Watch reports estimates that between 20 and 70 per cent of all CDM projects are non-additional. The NGO claims that large hydro-electric and coal-power projects, in particular, have repeatedly been shown to be business-as-usual.[68] Writing in 2014, the IPCC, citing an additionality study by Schneider which found a systematic lack of credible evidence in the relevant documentation in a sample of 93 CDM projects,[69] remarked that the issue of additionality continues to generate controversy.[70]

In another study, all 52 CDM projects registered in India by May 2006 were analysed for additionality testing.[71] Only half the projects had identified 'alternatives' in their certification applications. Only a third of the projects had undertaken an investment analysis. Independent sources of information had been used in only a third of the projects. About a fifth of the projects provided a common-practice analysis in adequate detail. The study found that DOE validators did not transparently evaluate the alleged barriers. Results from two of the projects which were studied in detail showed that 'additionality assessment by the CDM Executive Board varies; if the project developer can obfuscate the attractiveness of the project, it is more likely to pass'.[72]

Additionality can be undermined by 'leakage'. Leakage occurs when a CDM project displaces an unwanted activity, in whole or in part, to areas outside the boundaries of the project.[73] A project developer must provide leakage estimations for the proposed project for each year of the crediting period.[74] A study of stove-related CDM projects, aiming to reduce the use of wood fuel and attenuate the degradation of forests by supplying households within each project boundary with efficient stoves found that regions abutting the boundary used more wood fuel than they previously had, due to the lower cost and greater availability of the wood caused by the CDM-instigated drop in demand inside the boundary.[75] A DOE carrying out a verification to decide the number of CERs to issue to a stove project would have had to survey changes in wood fuel use not only inside the boundaries of the project but also in the zones surrounding it to check for

---

68 'Additionality and Baselines', <http://carbonmarketwatch.org/category/additionality-and-baselines/>.
69 See Lambert Schneider (2009), 'Assessing the Additionality of CDM Projects: Practical Experience and Lessons Learned', 9 *Climate Policy* 242.
70 IPCC (2014), *Climate Change 2014: Mitigation of Climate Change: Working Group III Contribution to the Fifth Assessment Report of the Intergovernmental Panel on Climate Change (Final Draft)*, ch. 13, p. 63.
71 Axel Michaelowa and Pallav Purohit (2007), *Additionality Determination of Indian CDM Projects: Can Indian CDM Project Developers Outwit the CDM Executive Board?*, Discussion Paper CDM-1.
72 Ibid., p. 1.
73 Kyoto Protocol (2005), *Decision 3/CMP.1*, para. 51.
74 CDM (2013), *Project Standard*, para. 50.
75 Prentiss Cox (2011), *Analysis of Cookstove Change-out Projects Seeking Carbon Credits*, p. 22.

leakage.[76] There is only so much verification that a DOE can do before costs blow out.

A notorious case of non-compliance with the additionality rule involved CDM projects set up to destroy HFC-23 gas, which is a by-product of the production of the (non-greenhouse) refrigerant gas HCFC-22. By mid-2014, there were nineteen HFC-23 CDM projects in existence—in China, India, Korea, Mexico, and Argentina—registered between 2005 and 2009 with emission reductions totalling 81.3 million CERs *per year* (offsetting the equivalent of Austria's total annual emissions; see Table 4.3 in Chapter 4).[77] For many years, manufacturers of HCFC-22 had been incentivized by the CDM to increase their production of HCFC-22, because this would lead to more HFC-23 being produced and thus to more CERs earned for capturing the gas and destroying it.[78] Where such overproduction occurred, the registered firms profited not from the additional HCFC-22 produced (whose price went down due to increased supply) but from the associated greenhouse gas.[79] CERs issued in such cases did not represent real emission reductions, so their use by buyers in Annex I countries did not represent a true offsetting of their emissions.[80] The CDM Executive Board, having finally become convinced by the protests of NGOs and others,[81] introduced safeguards in the applicable project-assessment methodology to prevent further abuse of the system. As a result of the reforms, no new projects can qualify for HFC-23 reduction, and the amount of HFC-23 destroyed at the plants that have been allowed to continue to qualify for CERs is tied to each plant's historical production levels.[82]

---

76 CDM Executive Board, *Methodology AMS-II.G* (Initial adoption 2008), p. 5.
77 The data may be retrieved from <http://cdm.unfccc.int/Projects/projsearch.html>, using the AM0001 methodology as the search term. (AM0001 is exclusive to HFC-23 reduction.) As the CDM Executive Board concedes, the HFC-23 projects account for the largest share of CERs issued to date: CDM Executive Board (2012), *Annual Report*, FCCC/KP/CMP/2012/3 (Part I), pp. 12–13.
78 An analysis of data from the 19 CDM-registered HFC-23 projects showed that two plants reduced HFC-23 generation when they were ineligible for crediting and increased HFC-23 generation once they could again claim credits for its destruction. One plant ceased HCFC-22 production when it was not allowed to generate further offset credits and resumed operation when it again became eligible to generate credits. The analysis also revealed that many plants produce exactly the amount of HCFC-22 and HFC-23 they are allowed to claim credits for, whereas production was lower or varied from year to year before offset credits were rewarded. See Environmental Investigation Agency (August 2010), *Ethically Bankrupt: World Bank Defense of the HFC-23 Scandal*, pp. 6–11.
79 Lambert Schneider (2009), 'Additionality of CDM Projects'.
80 Lambert Schneider (2011), 'Perverse Incentives under the CDM: An Evaluation of HFC-23 Destruction Projects', 11 *Climate Policy* 851.
81 See Carbon Market Watch <http://carbonmarketwatch.org/open-letter-to-patricia-espinosa-hfc-23-projects-undermining-credibility-of-unfccc-process/>.
82 The current version of the methodology fixes the baseline with reference to historical output in the period 2000 to 2004; see <http://cdm.unfccc.int/methodologies/PAmethodologies/approved>.

While additionality questions continue to affect the CDM, neither the Executive Board nor the Kyoto Protocol parties seem too concerned. The Board frequently reiterates its commitment to the 'environmental integrity' of the mechanism, while also emphasizing that additionality 'must go hand-in-hand with a strengthening of the CDM's efficiency and effectiveness'.[83] This alludes to a greater effort on the part of the Board to reduce costs and increase investment through 'simpler approaches [that] could be used to achieve the same result, where necessary by applying conservative default values or discount factors to ensure environmental integrity'.[84] The suggestion here is that 'the same result' could be achieved by making proof of additionality easier and by applying heavier conservativeness multipliers to make up for the relaxation in the rules. The Board has also promoted the theory of 'suppressed demand', which allows project developers setting up CDM projects in least-developed countries to assume levels of future development that rise above the current levels and lead to a higher baseline trajectory than a business-as-usual projection would allow. This increases the amount of emissions to be considered avoided through the operation of the project.[85]

6.3.2 Sustainable development: Limited knowledge

The CDM's additionality problems are relevant to both of the legal rules that make up the current climate change regime, namely accountable reporting and the general mitigation rule. States have allowed the CDM to be less than fully accountable by placing all the compliance burden solely on the Executive Board and by not developing a system of independent verification of its operations; also, by tolerating the creation of projects of low or non-existent additionality, states have undermined some of their own mitigation efforts by relying on offsets of doubtful value. Sustainable development does not present the same issues as additionality, mainly because this treaty requirement has no impact on mitigation. The Protocol's approach to sustainable development under the CDM shows, instead, a failure in the application of the rule on accountable reporting.

The Executive Board has said that it is 'paramount ... to maintain the prerogative of Parties to defin[e] their own criteria to be used' in determining whether a CDM project contributes to sustainable development.[86] The effect of this statement is twofold: it makes the definition of sustainable development relative to national context; and it relieves the Executive Board of responsibility for following up on this aspect of CDM projects. However, the Board's thesis is hollow. First, while it may be the case that different countries have different developmental priorities, it does not follow that a definition of sustainable development cannot be given in general terms. Second, the Executive Board has insisted that CDM projects do demonstrably deliver the benefit of sustainable development, and has produced

---

83 CDM Executive Board (2011), *Annual Report*, p. 9.
84 Ibid., p. 7.
85 Ibid., p. 7; and CDM Executive Board (2013), *Annual Report*, p. 17.
86 CDM Executive Board (2012), *Annual Report*, p. 12.

two studies to prove it.[87] Third, for its two studies, the Board has developed and applied a universal definition of sustainable development applicable to all projects in all countries. Fourth, scholars carrying out research into sustainable development have done much the same thing. Their definitions and the Board's are not substantially different. Fifth, the Board has produced a 'voluntary tool' for CDM project developers to use to describe the expected sustainable development benefits of their proposed projects in a 'consistent manner'. The Board developed the tool to get away from 'the current, free-form, sustainable development section of the project design document'.[88]

Thus, it is misleading of the Executive Board to suggest that the lack of an agreed operational definition of sustainable development, or even sovereignty concerns, explain why the responsibility for determining whether a CDM project contributes to sustainable development rests with the DNAs.[89] If either of these were true, the Board would not have produced its two studies. The real reason for the lacuna is simply that the Kyoto Protocol parties have not made an effort to settle on a definition of sustainable development and to follow through with a procedure on the assessment of the CDM's performance against this particular treaty requirement. This is a failure in accountable reporting which has led to responsibility for the CDM's sustainable development outcomes being cast adrift within the Protocol's framework.

NGOs and scholars have probed the question of the CDM's impact on sustainable development from the outside. According to Carbon Market Watch, the assessment process performed by the DNAs is usually perfunctory. The NGO claims that the interest of DNAs is to secure as many CDM projects as possible for their country because of the investment they bring. Sustainability benefits have no immediate financial value for a country, as only emission reductions lead to financial reward.[90] As a result, large-scale CDM projects (which comprise the majority), such as industrial-gas and hydro-electric projects, proliferate, yet they have few demonstrable sustainable development benefits, and some have serious negative impacts on local communities (such as displacement, job loss, and increased pollution).[91]

Schneider (whose study was cited by the IPCC in a comment referred to above) found no evidence in 93 CDM projects that host states had prioritized projects with a high sustainable development impact over those with little or no

---

87 CDM Executive Board (2011), *Benefits of the CDM 2011*; and CDM Executive Board (2012), *Benefits of the CDM 2012*.
88 CDM Executive Board (2012), *Annual Report*, p. 12.
89 CDM Executive Board (2012), *Benefits of the CDM 2012*, p. 13.
90 'Sustainable Development in the CDM', <http://carbonmarketwatch.org/category/sustainable-development/>. See also Michael Gillenwater and Stephen Seres (2011), *The Clean Development Mechanism: A Review of the First International Offset Program*, p. 30 ('Developing countries have few incentives to apply stringent criteria for sustainable development since they are effectively competing for CDM projects with other developing countries').
91 'Harmful CDM Projects', available at <http://carbonmarketwatch.org/category/project-campaigns/>.

such impact: 'This has resulted in a situation in which the CDM project portfolio is mainly determined by the economic attractiveness' of projects.[92] Disch studied 122 CDM projects in six countries to measure their 'development dividend'.[93] Only projects in one of these countries provided independent verification of sustainable development claims; for the rest, the whole subject was marginalized. Disch commented on the 'fundamental disconnect' in the international regime between the reporting of state emission reductions, on one hand, which is of high quality, and the assessment of sustainable development, on the other, for which no criteria have been developed.[94] Alexeew and colleagues evaluated 40 CDM projects for evidence of sustainable development as well as additionality.[95] The sample consisted of 31 small- and 9 large-scale projects. The relationship between the projects' additionality and sustainable development contribution was assessed to be a trade-off. Wind-power, small hydro-electric, and biomass projects provided relatively high sustainable development benefits but had a low probability of being additional. By contrast, energy-efficiency projects showed a high probability of additionality but resulted in low sustainability benefits relative to other types of project.[96]

The CDM Executive Board's two studies (the second was an update of the first) gave a much more upbeat account of the CDM's impact on sustainable development. However, methodologically the Board's two studies left a lot to be desired. They relied on the following two sources of evidence only: claims to sustainable development made by project developers in Project Design Documents;[97] and an on-line, non-anonymized, survey of CDM project developers which drew a response rate of 8.6 per cent.[98] The studies' conclusions were thus based on the pre-implementation claims of project developers and the post-implementation unverified claims of a small proportion of project developers who took the attributed survey. The Executive Board used the online responses as its basis for conclusions about the attainment of sustainable development impacts promised in the Project Design Documents of *all* projects covered by the two

---

92 Lambert Schneider (2007), *Is the CDM Fulfilling Its Environmental and Sustainable Development Objectives? An Evaluation of the CDM and Options for Improvement*, p. 47. See also Patrick Nussbaumer (2009), 'On the Contribution of Labelled Certified Emission Reductions to Sustainable Development: A Multi-Criteria Evaluation of CDM Projects', 37 (1) *Energy Policy* 91, p. 99.
93 David Disch (2010), 'A Comparative Analysis of the "Development Dividend" of Clean Development Mechanism Projects in Six Host Countries', 2 *Climate and Development* 50.
94 Ibid., p. 62. See also Christina Voigt (2009), 'The Deadlock of the Clean Development Mechanism: Caught between Sustainability, Environmental Integrity and Economic Efficiency', in *Climate Law and Developing Countries: Legal and Policy Changes for the World Economy*, ed. Benjamin J. Richardson et al. ('wider issues of long-term community benefit and localized environmental aspects tend to be ignored').
95 Johannes Alexeew et al. (2010), 'Additionality of CDM Projects and Their Contribution to Sustainable Development', p. 233.
96 Ibid., p. 244.
97 CDM Executive Board (2012), *Benefits of the CDM 2012*, p. 14.
98 Ibid., p. 21.

studies.⁹⁹ The exercise was tendentious and is best thought of as a publicity campaign rather than a true investigation.¹⁰⁰

At times, the Executive Board appears intent on drawing the attention of the Kyoto Protocol parties to the fact that the sustainable development rule has been neglected:

> There is a need to do more to make visible the sustainable development co-benefits of CDM projects. Sustainable development is part of the two-pronged objective of the CDM, as stated in Article 12 of the Kyoto Protocol, and yet it has been less visible than the emission impacts of projects. For the Board to ensure that the CDM makes an impact on sustainable development, it is crucial that DNAs set related criteria and ensure that they are met in the projects they approve.¹⁰¹

The Conference of the Parties to the Protocol passed the issue back to the Executive Board.¹⁰² No progress on this matter can be expected in the second commitment period. On the current evidence, we must conclude that after all these years we do not know for sure whether the CDM is making a positive overall contribution to sustainable development.

6.4 Mitigation of emissions through forest conservation

The FCCC's scheme for the protection and enhancement of forests in developing countries, known as REDD, is a vast, ambitious, and potentially crucial programme in climate change mitigation. Although it is only partially developed, enough of its elements are visible to allow for an analysis of the application of the climate change regime's legal rules in this area. REDD is a 'sink' programme for reducing greenhouse gas emissions from deforestation and forest degradation, which is why I have described it as a mitigation measure. It undoubtedly also offers adaptive possibilities, although they are not the focus of this book. The

---

99 Ibid., pp. 21–22.
100 As the Executive Board conceded, 'only positive contributions to sustainable development were assessed since project developers never state anything negative about their projects': ibid., p. 16. The Board had no way to check that projects were not *negatively* impacting sustainable development post-implementation: ibid., p. 16 ('No attempt was made to independently verify the sustainable development claims, so statements made may not reflect the actual delivery of the claimed sustainable development benefit', the Board concedes).
101 CDM Executive Board (2012), *Annual Report*, p. 7.
102 Kyoto Protocol (2013), *Decision 3/CMP.9, Guidance Relating to the Clean Development Mechanism*, FCCC/KP/CMP/2013/9/Add.1, para. 8 ('Requests the Executive Board to develop guiding tools to assist designated national authorities, upon the request of the host Party and on a voluntary basis, in monitoring the sustainable development benefits in its territory of clean development mechanism project activities and programmes of activities, recognizing that the use of such guiding tools is the prerogative of Parties').

Table 6.1 Greenhouse gas net emissions and removals (Mt CO_2 eq.) from the LULUCF sector of the main countries in the world's three largest tropical rainforests. Net removals are shown in italics. Note that the most recent available data is about a decade old.

Non-Annex I	1990	1995	2000	2005	2010
Brazil	812.7	1,937.8	1,328.3	1,329.0	–
Indonesia	197.7	164.1	821.3	–	–
D.R. Congo	–	*–176.8*	*–342.2*	*–178.8*	–

Source: FCCC, 'Greenhouse Gas Inventory Data: Detailed Data by Party' (online database), <http://unfccc.int/di/DetailedByParty.do>.

compliance challenges that REDD presents are large and unprecedented compared with other areas of the climate change regime.

The LULUCF sector in general accounts for about a quarter (10–12 Gt CO_2 eq.) of annual anthropogenic emissions. Deforestation is a leading source of emissions within the sector. The IPCC refers to forecasts indicating a decline in LULUCF emissions, largely due to decreasing deforestation rates and increased afforestation. Net LULUCF emissions are potentially less than half the 2010 level by 2050, and it is possible for the LULUCF sector to become a net sink before the end of the century.[103] A successful REDD mechanism is essential to this mitigation pathway. However, due to the uncertainties involved and the risks of reversal (not least due to the impacts of climate change), the IPCC finds that the long-term mitigation potential of forests, even with REDD operating, is unclear.[104]

REDD is meant to address not only deforestation but also common degrading activities such as selective logging, sub-canopy fires, and the collection of wood fuel. Limiting our focus to the world's three largest tropical rainforests, degradation is responsible for 20 per cent of total emissions in the Brazilian Amazon, two-thirds of forest-stock decrease in Indonesia (where the stock is decreasing at a rate of 6 per cent per year), and almost one-third of emissions in Africa.[105] As Table 6.1 indicates, LULUCF is reported as a net sink in the Congo rainforest, but this could be reversed as the region develops.

6.5 REDD: State obligations under the Framework Convention

The text of the FCCC supplies no helpful foundation on which to build a scheme to protect and enhance the world's forests as carbon sinks. Article 4.1.d binds all parties to a general commitment that they will work towards sustainable management, conservation, and enhancement of forests. Following this, the subject of forests is not taken up again in the treaty text. The substance and nature

---

103 IPCC (2014), *Climate Change 2014: Mitigation of Climate Change: Working Group III Contribution to the Fifth Assessment Report of the Intergovernmental Panel on Climate Change: Summary for Policymakers*, p. 25.
104 IPCC (2014), *5AR WG3 (Final Draft)*, Chapter 11, pp. 44–45.
105 Daniel Murdiyarso et al. (2012), 'Some Lessons Learned from the First Generation of REDD+ Activities', 4 *Current Opinion in Environmental Sustainability* 678, p. 683.

154   *'Optional' mitigation mechanisms*

of legal obligations that REDD gives rise to must be derived from decisions of the plenary of the parties.[106]

As for the Kyoto Protocol, with its narrow focus on developed countries, predictably it makes no advance on the protection of the world's forests. Article 2 of the treaty requires Annex I parties to protect and enhance their sinks and reservoirs of greenhouse gases and to promote practices of sustainable forest management. Emissions from deforestation are included in Annex I parties' emission inventories (in the LULUCF sector under FCCC reporting) and, for those states that are parties to the Protocol, they also count towards their commitment-period targets (by virtue of article 3.3 of the Protocol). In this way, there is an indirect limit on deforestation in the Annex I region, but there is no specific limit. The only Annex I party with a serious deforestation problem is Russia.[107]

In the above sense, then, the international climate change regime already controls deforestation emissions from Annex I parties. Non-Annex I parties, by contrast, are not regulated at all, and it is primarily these countries that are responsible for present-day deforestation.[108] The closest the Protocol comes to engaging developing countries in sink activities is through the CDM, which is insufficient, in view of the scale of the problem.[109]

The development of a REDD mechanism was first endorsed by the FCCC parties in 2007.[110] The task was incorporated into a broad, initially two-year, negotiating track on long-term cooperative action outlined in the Bali Action Plan.[111] The

---

106   On REDD's tenuous normative status, see Annecoos Wiersema (2013), *Climate Change, Forests, and International Law: REDD's Descent into Irrelevance*, p. 27 ('REDD … has been developed under the auspices of the UNFCCC, but no formal treaty or binding agreement has been negotiated to implement it').

107   On Russia's deforestation emissions, see Forest Europe, UN Economic Commission for Europe, and Food and Agriculture Organization (2011), *State of Europe's Forests 2011: Status and Trends in Sustainable Forest Management in Europe*, p. 20. Russia is not the only Annex I party with a deforestation problem, even if Russia's is the most serious. Deforestation in Australia, for example, continues at significant levels. Over the period 2008–2010, the country's emissions from deforestation totalled 133 Mt CO_2 eq. (see Expert Review Team (2013), *Report of the Individual Review of the Annual Submission of Australia Submitted in 2012*, FCCC/ARR/2012/AUS, p. 5). For more on the state of forests in Annex I countries, see Food and Agriculture Organization (2010), *Global Forest Resources Assessment 2010: Main Report*, pp. xvi–xvii; and Christopher Potter et al. (2008), 'Storage of Carbon in U.S. Forests Predicted from Satellite Data, Ecosystem Modeling, and Inventory Summaries', 90 *Climatic Change* 269.

108   For deforestation rates in REDD countries for the period 2000–2010, grouped into high (>100,000 ha/year), medium (50,000–100,000 ha/year), and low (<50,000 ha/year) rates, see Daniel Murdiyarso et al. (2012), 'Lessons Learned', p. 679.

109   CERs from CDM afforestation and reforestation projects can be used only as temporary offsets, as noted earlier in this chapter. They have thus played a very minor role in the CDM.

110   For an overview of the history REDD, see Federica Bietta (2010), 'From the Hague to Copenhagen: Why It Failed Then and Why It Could Be Different', in *Deforestation and Climate Change: Reducing Carbon Emissions from Deforestation and Forest Degradation*, ed. Valentina Bosetti and Ruben Lubowski.

111   FCCC (2007), *Decision 1/CP.13, Bali Action Plan*, FCCC/CP/2007/6/Add.1, para. 1.

plan classifies REDD as a mitigation measure. It urges parties to consider 'Policy approaches and positive incentives on issues relating to reducing emissions from deforestation and forest degradation in developing countries' (the basic form of REDD), as well as forest conservation, sustainable management of forests, and enhancement of forest-carbon stocks (sometimes referred to as REDD-plus).[112] The language ('positive incentives') suggests that developing countries would be paid to carry out mitigation actions under the REDD scheme, and that REDD would develop into a form of international financial transfer earmarked for forest-based mitigation.

A second decision at Bali gave FCCC parties a green light to proceed with REDD-related actions of their own, on a voluntary basis.[113] The intention was to lay the groundwork for the scheme in developing countries wishing to participate in the precursory stage, even as the negotiations on REDD continued. The groundwork has come to include capacity-building financial support to developing countries, the carrying out of REDD 'demonstration' activities, tentative efforts in monitoring and reporting results, and so on.[114] The second Bali decision additionally tasked the SBSTA to commence a programme of technical work on methods for determining forest reference emission levels (that is, business-as-usual trajectories, or baselines, against which progress would be measured) and assessing change in forest-carbon stocks for the demonstration of emission reductions from deforestation and forest degradation.[115]

At the 2009 Copenhagen conference, the FCCC called on non-Annex I parties to establish 'robust and transparent' national forest-monitoring systems that use a combination of remote sensing and ground-based forest-carbon inventory methods for estimating 'anthropogenic forest-related greenhouse gas emissions by sources and removals by sinks, forest carbon stocks and forest area changes'; these were required to ensure 'estimates that are transparent, consistent, as far as possible accurate, and that reduce uncertainties'.[116] Developing countries implementing REDD activities were requested by this decision to apply the IPCC guidelines on accounting for forestry emissions.[117]

The next notable decision by the FCCC parties on REDD was taken at the Cancun conference in 2010. The plenary called on participating non-Annex I parties to proceed with the implementation of specified elements at the state level: a national strategy and action plan for REDD implementation (produced with the monetary and technical support, wherever needed, of developed countries);

---

112 Ibid., para. 1.b.iii.
113 FCCC (2007), *Decision 2/CP.13, Reducing Emissions from Deforestation in Developing Countries: Approaches to Stimulate Action*, FCCC/CP/2007/6/Add.1, paras 3–4.
114 Ibid., para. 4.
115 Ibid., para. 7.a.
116 FCCC (2009), *Decision 4/CP.15, Methodological Guidance for Activities Relating to Reducing Emissions from Deforestation and Forest Degradation*, FCCC/CP/2009/11/Add.1, para. 1.
117 See IPCC (2003), *Good Practice Guidance for Land Use, Land-Use Change and Forestry*.

156 *'Optional' mitigation mechanisms*

a national forest reference emission level;[118] a 'robust and transparent' national system for the monitoring and reporting of REDD activities; and a reporting system through which to report on compliance with the several procedural and substantive 'safeguards' itemized in this decision.[119] As land-tenure issues are central to any scheme that seeks to distribute benefits for the conservation of forested land, the 2010 decision called on developing countries to clarify the legal status of land destined for REDD projects.[120] Some of this repeated earlier decisions, but the safeguards element was new.

The main REDD safeguards are as follows. REDD actions must complement or be consistent with the 'objectives' of national forest programmes and 'relevant' international conventions and agreements. 'Transparent and effective' national forest-governance structures must be in place that 'take into account' national legislation and 'sovereignty'. The knowledge and rights of indigenous people and members of local communities must be 'respected' by taking into account 'relevant international obligations [and] national circumstances and laws'. Stakeholders in REDD actions, in particular indigenous people and local communities, must enjoy full and effective participation. REDD actions must be consistent with the conservation of natural forests and biological diversity and must not be used for the conversion of natural forests. They must be used to incentivize the protection and conservation of natural forests and their ecosystem services and to enhance 'other social and environmental benefits'. A final safeguard is that REDD actions must address the risk of 'reversals' and reduce the displacement or leakage of emissions.[121]

While the very term 'safeguards' suggests that a compliance system might be developed around them, the Cancun decision went no further than to require that the safeguards 'should be promoted and supported' by the FCCC parties.

The 2010 decision gave permission, as an interim measure, for the development of *subnational* forest reference levels.[122] This was because most countries with tropical forests did not as yet have the capacity to produce a national forest reference emission level. The decision thus affirmed the FCCC's plan that REDD would be implemented in phases, beginning with demonstration activities. National forest reference emission levels would be deferred in those cases where only subnational reference levels are feasible. Eventually these actions would

---

118 Forest reference baselines are developed by taking into account historical data and adjusting for national circumstances: FCCC Subsidiary Body for Scientific and Technical Advice (27 November 2011), *Report on the Expert Meeting on Forest Reference Emission Levels and Forest Reference Levels for Implementation of REDD-Plus Activities*, FCCC/SBSTA/2011/INF.18, para. 16. Baselines are not fixed but must be adjusted or recalculated over time to encompass changes in management, government policy, and patterns and causes of land-use change; see Nicole R. Virgilio et al. (2010), *Reducing Emissions from Deforestation and Degradation (REDD): A Casebook of on-the-Ground Experience*, p. 4.
119 FCCC (2010), *Decision 1/CP.16, The Cancun Agreements: Outcome of the Work of the Ad Hoc Working Group on Long-Term Cooperative Action under the Convention*, FCCC/CP/2010/7/Add.1, paras 68–79.
120 Ibid., para. 72.
121 Ibid., Appendix I, para. 2.
122 Ibid., Appendix I, para. 71.b.

mature into countrywide forest-conservation measures that would be reported on, verified, and somehow compensated financially.[123]

In 2011, the FCCC conference in Durban continued the evolution of REDD, settling some technical details about how states are to calculate their forest-related emissions, and launching a process to explore REDD's funding mechanism. The conference also called on participating non-Annex I parties to submit information on the development of their forest reference emission levels.[124] It said that the levels should be transparently set and should maintain consistency with estimates of anthropogenic emissions and removals as contained in each country's greenhouse gas inventory. A process for the technical assessment of proposed reference levels was established.[125] The conference reiterated that subnational reference levels are only an interim stage along the way to national reference levels, but did not say when they might be phased out.[126] The decision also reiterated that participation in the REDD scheme is voluntary for developing countries.[127] The FCCC's pronouncements on REDD have been ambiguous on whether Annex I financial support for REDD is also to be voluntary.

The FCCC parties have not settled on a definition of 'forest' for REDD purposes. Because most countries already have a definition of 'forest' in their domestic legislation, the FCCC decided that each participating non-Annex I party would determine its forest reference level using its own definition, as long as the same definition is used by the party in other international reporting on forests.[128]

The question of how REDD would be financed was addressed again in a 2013 decision.[129] The main point was that while 'results-based finance' provided to developing countries for REDD activities (viz. finance for confirmed emission reductions) could come from a variety of sources (public and private, bilateral and multilateral), it was to be new, additional, and predictable.[130] Developing countries undertaking REDD preparatory and partial- or full-implementation activities would qualify for financial support only if their actions are fully reported on and verified in accordance with two other decisions issued at the same conference.[131]

---

123 Ibid., para. 73. National coverage of REDD is needed to ameliorate the issue of emission displacement (leakage), a major drawback of subnational approaches; see IPCC (2014), *5AR WG3 (Final Draft)*, Chapter 11, p. 72.
124 FCCC (2011), *Decision 12/CP.17, Guidance on Systems for Providing Information on How Safeguards Are Addressed and Respected and Modalities Relating to Forest Reference Emission Levels and Forest Reference Levels as Referred to in Decision 1/CP.16, Appendix I*, FCCC/CP/2011/9/Add.2, para. 9.
125 Ibid., para. 15.
126 Ibid., para. 11.
127 Ibid., para. 13.
128 Ibid., Annex.
129 FCCC (2013), *Decision 9/CP.19, Work Programme on Results-Based Finance to Progress the Full Implementation of the Activities Referred to in Decision 1/CP.16, Paragraph 70*, FCCC/CP/2013/10/Add.1.
130 Ibid., para. 1.
131 Ibid., para. 3. The two other decisions are FCCC (2013), *Decision 13/CP.19, Guidelines and Procedures for the Technical Assessment of Submissions from Parties on Proposed Forest Reference Emission Levels and/or Forest Reference Levels*,

A further condition for financial support was that a developing country applying for 'results-based' payments is to provide a recent summary of information about how each of the REDD safeguards has been addressed by the country.[132]

In the 2013–2020 period, REDD emissions and removals are to be quantified in the context of greenhouse gas inventories which are to be submitted by non-Annex I parties in their biennial reports.[133] For those parties seeking result-based payments for REDD activities, the information on forest reference emission levels and emission-reduction activities claimed in the biennial reports is to be analysed by a 'technical team of experts', composed of LULUCF experts selected from the FCCC's Roster of Experts, which is to publish the results of its analysis and identify areas for improvement.[134] A separate decision provides information on the scope of the technical assessment. The assessment includes checking for consistency with emission/removal estimates of REDD activities, how historical data have been used, transparency, completeness, accuracy, consistency of the definition of 'forest' with definitions used by the non-Annex I party in other reporting, clarification of assumptions about future changes in domestic policies included in the reference levels, and carbon pools and gases included or omitted.[135]

The steady improvement in rules of reporting that I have discussed in other parts of the book will undoubtedly also benefit the REDD programme. However, a mere process that involves a technical team of experts in discreetly analysing update reports with no further compliance elements is not sufficient for 'accountable reporting' for REDD, considering the potential profits involved and the risks of misinformation that REDD gives rise to (a matter which I will address in the next section).[136]

From the substantive perspective, the issue of payment for REDD activities is far from settled. No matter how much the reporting systems on REDD are improved, there will be no substantial mitigation effort before a payment method is agreed to. The 2013 FCCC conference instructed the Standing Committee on Finance to urgently focus on issues related to REDD finance.[137]

A financing mechanism for REDD could be organized as a centralized fund that donor countries contribute to, and out of which payments are made to countries that reduce emissions. Alternatively, in a market-based approach, REDD credits

---

FCCC/CP/2013/10/Add.1 and FCCC (2013), *Decision 14/CP.19, Modalities for Measuring, Reporting and Verifying*, FCCC/CP/2013/10/Add.1.

132 FCCC (2013), *Decision 9/CP.19*, para. 4.
133 FCCC (2013), *Decision 14/CP.19*, paras 6–7.
134 Ibid., paras 7–11.
135 FCCC (2013), *Decision 13/CP.19*, Annex. See also FCCC (2013), *Decision 11/CP.19, Modalities for National Forest Monitoring Systems*, FCCC/CP/2013/10/Add.1, para. 4.
136 The guidelines and procedures for technical assessment do not go beyond a 'facilitative, non-intrusive, technical exchange of information'; the assessment team 'shall refrain from making any judgment on domestic policies taken into account in the construction of forest reference emission levels' (FCCC (2013), *Decision 13/CP.19*, Annex). There is no provision about what to do in cases of disagreement between the technical team and the state under assessment.
137 FCCC (2013), *Decision 9/CP.19*, para. 20.

would be generated on the CDM model (as offsets) and traded in compliance markets.[138] The problem with the latter approach is that the FCCC has not imposed any binding emission-reduction targets on states, as the Kyoto Protocol has. Because the FCCC targets for 2013–2020 are voluntary, it is doubtful that FCCC parties would go as far as to buy offsets from the REDD mechanism to comply with their voluntary targets, especially as these offsets would not be recognized under Protocol accounting rules.

6.6 Compliance issues with REDD's implementation

FCCC parties are collectively under an obligation 'to slow, halt and reverse forest cover and carbon loss ... consistent with the ultimate objective of the Convention'.[139] This is not a separate rule of law but an aspect of the FCCC's general mitigation rule. REDD is essentially the only mechanism devised by the climate change regime to enable mitigation of emissions from forests. According to the analysis I have presented in this book, the mechanism can be critiqued through the examination of two fundamental questions. First, as a matter of substance, have the FCCC parties utilized REDD to meet their mitigation obligations under the general migration rule? Second, as a matter of procedure, have the parties complied with the requirements of the accountable reporting rule as applied to REDD? These same types of question were considered in connection with the CDM in the first half of this chapter. The critical differences between the two examinations are, first, that REDD, which is meant as a permanent programme, is far more ambitious than the CDM and implies political and cultural reforms over vast geographical areas and under heavy time constraints. Second, REDD is underdeveloped, so the substantive question is not as easily answered as the procedural one.

A national approach to REDD requires the determination of a national forest reference emission level and the establishment of a national information system on REDD-related forest work and the ongoing emissions from the state's forests. Because few developing countries in the tropics have funds for regular forest

---

138 Or it could be a combination of the two: FCCC (2011), *Decision 2/CP.17, Outcome of the Work of the Ad Hoc Working Group on Long-Term Cooperative Action under the Convention*, FCCC/CP/2011/9/Add.1, paras 66–67. Isenberg and Potvin argue that the financial approach for REDD should involve two tracks: a 'market track' serving as a mitigation option for developed countries subject to compliance markets, and a non-offset 'fund track' serving as a mitigation option for developing countries: Jordan Isenberg and Catherine Potvin (2010), 'Financing REDD in Developing Countries: A Supply and Demand Analysis', 10 (2) *Climate Policy* 216, p. 216. The market track would subsume all those activities in which a differential in emissions or carbon stocks could be measured relatively accurately, such as reduction in emissions from deforestation and forest degradation as well as increments in forest-carbon stocks. The fund track would allow the inclusion of activities such as sustainable forest management and conservation of forests, whose impact on emissions is not as accurately measurable: ibid., p. 227.
139 FCCC (2013), *Decision 13/CP.19*.

inventories, even basic information on the extent and condition of forests is often outdated or non-existent.[140]

In 2005, the final instalment of the compilation and synthesis report on the first national communications of non-Annex I parties (the only such report to be completed to date) observed that, in general, LULUCF emission data was lacking or was highly uncertain. Developing countries lacked relevant technical capacity, in particular the mechanisms needed to systematically collect data on forests, undertake field studies to validate emission factors, carry out surveys to reduce uncertainties in activity data, improve the use of methodologies to determine forest area, or improve their institutional capacity in all of these matters.[141] The report is about a decade old, however little has changed in the meantime.

In 2012, Romijn and colleagues reported on a study of 99 tropical developing countries which set out to assess the status of their national monitoring capacities.[142] Forty-nine countries, mostly in Africa, were found to have 'very large' capacity gaps, according to the metric employed in the study, while only four countries had a 'very small' capacity gap. The majority of countries lacked capacity, at technical, political, and institutional levels, to implement a complete and accurate national monitoring system to estimate forest-area change and to attribute emissions to these changes using the IPCC's guidelines on LULUCF reporting.[143] Countries with a 'small' capacity gap, including Mexico and India, could measure forest-area change and perform a regular national forest inventory on growing stock and forest biomass. In the Congo region, by contrast, there was low engagement in the REDD process and a large gap in remote-sensing capacity. Satellite remote sensing is necessary for monitoring deforestation because it is the only practical means to cover large areas of forest for national-level monitoring.[144] According to the Romijn study, most countries in Asia and South America had a small to medium capacity gap. Although generally engaged in the REDD process, their capacity to estimate changes in carbon pools was low.[145] Even allowing for an expansion of satellite remote sensing, cloud cover and large variations in altitude in mountainous countries such as Ecuador and Peru impose limits on the accuracy of information derived from satellite images.[146] The study concluded that there

---

140 FAO and ITTO (2011), *The State of Forests in the Amazon Basin, Congo Basin and Southeast Asia: A Report Prepared for the Summit of the Three Rainforest Basins, Brazzaville, Republic of Congo, 31 May – 3 June, 2011*, p. 3.
141 FCCC Subsidiary Body for Implementation (25 October 2005), *Sixth Compilation and Synthesis of Initial National Communications from Non-Annex I Parties; Addendum: Inventories of Anthropogenic Emissions by Sources and Removals by Sinks of Greenhouse Gases*, FCCC/SBI/2005/18/Add.2, para. 90.
142 Erika Romijn et al. (2012), 'Assessing Capacities of Non-Annex I Countries for National Forest Monitoring in the Context of REDD+', 19–20 *Environmental Science and Policy* 33.
143 Ibid., p. 39.
144 Ibid., p. 34.
145 Ibid., p. 39.
146 Ibid., p. 41. On the difficulties in interpreting remote-sensing data for tropical forests, see also Nicole R. Virgilio et al. (2010), *REDD Casebook*, p. 4 ('remote sensing ... is still not very accurate in assessing change in the carbon content of forests. To determine

must be 'realistic expectations of what developing countries can reasonably do in this area', and that a large capacity-building effort would be needed before REDD could operate as a mechanism that attracts 'results-based finance'.[147]

Another study from 2012 surveyed 17 REDD demonstration sites in Latin America, Africa, and Southeast Asia.[148] It found that 53 per cent of the projects used site-specific or country-specific allometric equations for the assessment of above-ground biomass, as would be required for a Tier 2 (medium-accuracy) IPCC approach, whereas 47 per cent used generalized equations for the tropics. Only one-quarter of the project teams could estimate below-ground biomass using recognized methods. For litter and soil-carbon pools, the study found that most projects were either using IPCC default values or ignored these pools.[149] In relation to clarity of tenure rights over forests, trees, and forest carbon, the REDD projects in the countries studied, including in Brazil, Cameroon, Tanzania, Indonesia, and Vietnam, were being implemented in a tenure context where 'contestation and overlapping claims are rife'.[150]

Forestry in developing countries attracts much illegal activity, which will be a major obstacle to the establishment of REDD in some countries.[151] In addition to vested interests,[152] or regular pressures of economic development,[153] profit-seeking behaviour in the context of weak law enforcement will continue to support intense deforestation. The Environmental Investigation Agency (an NGO) reports that flagrant violations of forestry law in Laos are the norm due to a well-established trade in illegal logging between Laos and Vietnam.[154] Vietnam's booming wood-processing and furniture industries depend on that trade. The EIA alleges that, within Laos, the profit from illicit logging goes to government officials, the military, and businessmen. Both Laos and Vietnam have introduced policies to

---

carbon content, it is necessary, among other things, to measure the circumference of trees'); and Kamel Soudani and Christophe François (13 February 2014), 'A Green Illusion', 506 *Nature* 165.
147 Erika Romijn et al. (2012), 'Assessing Capacities for National Forest Monitoring', p. 44.
148 Daniel Murdiyarso et al. (2012), 'Lessons Learned'.
149 Ibid., p. 682.
150 Ibid., p. 681.
151 Corruption is not limited to developing countries, of course. In 2011, Russia supplied around 25 per cent of roundwood and sawnwood exports to China, much of it illegally harvested: D. Y. Smirnov et al. (2013), *Illegal Logging in the Russian Far East: Global Demand and Taiga Destruction*, p. 9. Corruption within Russia's public agencies plays a role in the low levels of prosecution of illegal logging: ibid., p. 18.
152 Daniel Murdiyarso et al. (2012), 'Lessons Learned', pp. 679–680 ('deforestation in REDD countries (such as Brazil, Cameroon, Indonesia, Vietnam, Bolivia, and Mozambique) is driven by agriculture and mining industries, which are often well connected to existing bureaucratic structures').
153 Ibid., p. 680 ('In Vietnam, the national program for expansion of hydropower plants and infrastructure to meet national development goals is considered to contribute to a rapid increase of deforestation and forest degradation').
154 Environmental Investigation Agency (2011), *Crossroads: The Illicit Timber Trade between Laos and Vietnam*.

exclude illegally logged timber, yet the flow continues unabated, according to the NGO,[155] which cites studies that estimate that nearly 50 per cent of Vietnam's timber imports come from illegal sources.[156]

A workable system for REDD also seems a long way off in Kenya, where corruption in forest management is described as widespread.[157] Problems in Kenya also stem from contested land rights and forced evictions of people from forests. Indeed, 'Many countries lack the institutional capacity and legal safeguards to ensure that a centralized REDD regime would equitably allocate incentives to local actors.'[158] The UN-REDD Programme[159] has expressed concern that all systems of managing finance for REDD activities risk failure in a context of lack of transparency and poor accountability in the forestry sector: 'risks exist that funding for REDD activities and land use planning will favour powerful interests and political elites, and that the right to free, prior, and informed consent of those affected may not be effectively implemented'.[160] In Nepal, 'poor governance, corruption, and institutional issues pose challenges to the successful implementation of REDD'.[161] In several countries that are implementing REDD, deliberately erroneous monitoring, inaccurate or inflated accomplishment reports, and resistance to audits are the norm in the forestry sector.[162]

Wiersema has argued that REDD's safeguards are subject to minimal international oversight. The international community, she writes, has created 'an instrument which [is] country-driven and voluntary with limited international oversight and even limited international guidance ... limiting the role of international law in REDD'.[163] As in the CDM's case, REDD will be required to work according to the letter of its design if FCCC parties are to claim that they are discharging their obligations through the scheme, especially if REDD is one day to be used for offsetting purposes. There is, moreover, still no indication that REDD finance will flow at a rate sufficient to create significant emission reductions, even if the reporting and verification problems were solved.[164] To illustrate the magnitude of the flow problem, Neeff and Ascui have argued that even a total redirection to

---

155 Ibid., p. 1.
156 Ibid., p. 19.
157 UN-REDD Programme (2013), *Sharing National Experiences in Strengthening Transparency, Accountability and Integrity for REDD+*, p. 3.
158 Nicole R. Virgilio et al. (2010), *REDD Casebook*, p. 6.
159 Much of the money for REDD's precursory stage is contributed by the UN-REDD Programme, which is not an organ of the FCCC. UN-REDD was created following the Bali conference of the FCCC parties and is made up of the FAO, UNDP, and UNEP. Grants are made by the Programme to countries to set up their national REDD programmes in preparation for full implementation.
160 UN-REDD Programme (2013), *Strengthening Transparency*, p. 4.
161 Ibid., p. 11.
162 Ibid., p. 8.
163 Annecoos Wiersema (2013), *REDD's Descent into Irrelevance*, pp. 48–49 and 46.
164 Till Neeff and Francisco Ascui (2009), 'Lessons from Carbon Markets for Designing an Effective REDD Architecture', 9 *Climate Policy* 306, p. 309.

REDD of all existing GEF funds or all existing official development assistance would still be insufficient to channel enough capital to REDD.[165]

6.7 Conclusion: Rules versus action

The CDM and REDD are weakly regulated in critical performance areas; they favour rapid implementation and expansion of their operations; and they diffuse responsibility among multiple actors to such an extent that accountability is often lost. State compliance with the rules for these two schemes is difficult to assess, with the result that mitigation outcomes are (or, in the case of REDD, will be) doubtful. Transparency and environmental integrity compete for attention against the higher-profile need of keeping these two mitigation mechanisms afloat in a context of state refusal to take on adequate and binding mitigation targets. The frustration is evident in a plea by the CDM Executive Board:

> it is the current low level of demand for CERs and resulting low level of activity that is of gravest concern to the Board. This mature and now well-functioning mechanism, which has proven its ability to achieve emission reductions at a scale that can contribute significantly to the global mitigation effort and which has proven that it can attract substantial capital (public and private) and technology to developing countries, is at risk. Capacity built by project developers, DOEs, designated national authorities (DNAs) and within the secretariat could be lost, projects may be discontinued and low-cost mitigation opportunities missed.[166]

The REDD mechanism will raise compliance issues far greater than the FCCC and Kyoto Protocol have ever had to deal with. Since REDD is still a long way from being fully operational, this chapter has only tentatively attempted to assess state compliance in actual practice. It is already evident, however, that the FCCC's accountable reporting rule will not be implemented in the context of REDD at the required legal standard for many years to come. Streck and Lin have argued that, in order to create an effective REDD mechanism, strong international institutions for quality assurance are needed, yet their reference to the CDM as a kind of standard does not take account of the CDM's own lack of rigour, according to the arguments I have presented in this chapter. They are probably right, nevertheless, in thinking that the CDM is the more manageable of the two:

---

165 Ibid., p. 310. High implementation costs and financial uncertainty could themselves cause REDD to fail: Robin B. Matthews, Meine van Noordwijk, Eric Lambin, Patrick Meyfroidt, Joyeeta Gupta, Louis Verchot, Kristell Hergoualc'h, and Edzo Veldkamp (2014), 'Implementing REDD: Evidence on Governance, Evaluation and Impacts from the REDD-ALERT Project', 19(6) *Mitigation and Adaptation Strategies for Global Change* 907–925, p. 920.
166 CDM Executive Board (2013), *Annual Report*, p. 7.

a governance structure that relies on the delegation of power and control to an international review mechanism should be created to administer the REDD mechanism when it emerges. In this regard, parties to the UNFCCC should contemplate the delegation of certain elements of power to an internationally-established market regulator. ... Experience with the CDM should inform the development of better accountability and due process features for the REDD mechanism.[167]

---

[167] Charlotte Streck and Jolene Lin (2009), 'Private Sector Involvement in International Carbon Finance', pp. 73, 96.

7 Compliance lessons from the climate change regime

In this book I have surveyed climate change law as well as state compliance with that law in the period from 1997, when the FCCC went into force, to 2014, when reporting and other procedures relating to the Kyoto Protocol's first commitment period (2008–2012) were being completed.

Climate change is a global environmental problem that for the most part does not seem to be regarded as a very *pressing* problem. Some say it is the greatest challenge of our time, but few act as if it is. The problem is complex and long-lived and is not directly experienced—like, say, a major oil spill, to which we invariably respond with urgency until it is under control or goes away. Climate change is potentially much more destructive than any oil spill could be, and while this is understood by practically everyone engaged with climate change issues, in the course of a typical year climate change will not make the national headlines on more than a few days, say around the time of the FCCC Conference of the Parties or if an extreme and destructive weather event occurs.

I make this observation so as to avoid the artificial mystery that would result from juxtaposing a supposedly highly urgent problem with an apparently modest state response to it. The general attitude among states at the periodic climate change negotiations is that, while there must be *some* progress in the negotiations from year to year, the issues that cannot be resolved in the current year may be carried over to the next. Notoriously, a solution to the grand issue of a new agreement for 'long-term cooperative action' was postponed by *several years* when the negotiations produced no new agreement in Copenhagen in 2009. In this gradualist, Lamarckian context of 'some action' against climate change, we would expect to find 'some compliance' with the growing corpus of regime rules. It would be surprising indeed to discover a practice of zero compliance, or a practice of perfect compliance. What we expect to find in a modestly ambitious regime is a modest effort to implement what the regime demands.

In developing the argument in this book, I began with the link between law and compliance. Lawmaking is a kind of rulemaking. Rules may be divided into rules of law and other rules. In a treaty context, rules of law are those binding obligations in the treaty text usually made explicit through mandatory language, but sometimes necessarily implied. A state may claim compliance with a treaty when it fulfils (or is fulfilling) its binding obligations.

Because of the law-compliance link, an analysis of the compliance of states with the climate change treaties necessitates the prior identification and clarification of the rules of treaty law with which the state parties must comply. These should be distinguished from other rules for which compliance is not obligatory as a matter of law. I have argued that two rules of law stand out in the climate change regime. They are the accountable reporting (or transparency) rule and the general mitigation (or prevention) rule. I will return to review these below.

An observation that has been important to my argument is that a rule of law can take different forms. It can be directed to the conduct of individual states, or to that of the community of states. In the latter case, compliance is an attribute of collective conduct. A state alone cannot, logically, comply with a rule that is communal in form. For a state to have obligations derived from a communal rule, the compliance burden of that rule must be partitioned and shared among the states. Following this, each state will carry the burden that has been assigned to it and will be individually responsible for meeting it. If a communal rule implies from its context that states must 'individualize' the rule to give it effect, then this aspect of the rule requires specific conduct at the state level, even before a burden-sharing agreement is struck. A state may fail to comply with this implied legal obligation if, for example, the state acts to undermine attempts by the community of states to individualize the rule.

It might seem that the most effective form of law is one that specifies the conduct required of each state. This is true if all things are equal regarding normativity. A law that is merely 'technocratic' (that aims to force a result without any basis in a widely shared and agreed upon principle) will lack normativity and will not be as effective as a law that has moral compulsion. It is thus possible for an individualized law (such as the mitigation rule covering the Protocol's second commitment period) to be no more effective than a communal law (such as the FCCC's general mitigation law which controls the pledge period running parallel to the Protocol's second commitment period), in a case where the normative hold of the former is contested whereas that of the latter has long been accepted.

Where a communal rule is not complied with, states collectively will be in non-compliance, but not states individually. To avoid this legally disorienting condition—which is a kind of legal holding pattern—and in order to give proper effect to the rule, state parties to a treaty have no option but to transform a communal rule into state-level obligations, as already noted. A rule that implies a time limit for action (as the FCCC's prevention rule does) additionally implies that, for as long as states delay the step of individualization, they will be in breach of the law they have created. The implied obligation to individualize responsibility itself takes a communal form (all states must come to agree on how to partition the collective obligation), although it also has implications for conduct at the state level: a state that does not cooperate in good faith with other states to particularize the effort needed to meet the communal rule's objective will be in breach of the implied obligation.

Certain rules developed under a treaty regime may be opted into by states. Compliance with the rules of 'optional' mechanisms is of importance to those states that have chosen to participate in them. But it may also be an issue that affects all state parties. The laws of a regime (such as the rules on accountable reporting and prevention in the case of the climate change regime) logically extend to all corners of the regime. The creation of an optional mechanism is no excuse to flout them. In the climate change regime, certain optional mechanisms, such as the CDM and REDD, are mitigation mechanisms. 'Compliance' in the context of these mechanisms cannot be reduced to conformity with their technical rules only. The regime's foundational rules apply to them too, and they supply the criteria of a deeper evaluation of the mechanisms' operations, as well as insights into the true compliance of states with the regime's rules.

Next in my argument was the identification of normative international climate law. At the risk of sounding repetitive, I will briefly reiterate my main findings. In the category of procedural law, one rule stands out. Accountable reporting of state greenhouse gas emissions is a binding rule of climate change treaty law at the state level. The rule is universally accepted by states and is widely complied with. The extent of its implementation is expanding within the FCCC, with a clear tendency eventually to include all parties under a single model of implementation to ensure transparency of action in climate change matters. I have suggested that accountable reporting in the climate change regime is gradually hardening into a customary rule of international law. The basis for such an argument is that there has been no serious dissent from the rule, and the rule is constantly reaffirmed by state and non-state actors alike. However, some disagreement still exists about the extent of its application to non-Annex I parties, as evidenced by the distinction between the IAR and ICA procedures. The disagreement can be traced to the continuing unequal distribution of resources and capacity around the world, which affects the reporting ability of many states, as well as to the politics surrounding state 'leadership' on climate action. Nevertheless, it should be emphasized that there is no evidence of disagreement about the necessity for, and normative foundations of, accountable reporting. A hesitation in implementation, rather than a resistance in principle, is what is denying ICL status to accountable reporting. My examination of reporting practice in the context of the CDM and REDD, in particular, has shown that states have made no real effort to extend the rule on accountable reporting to these areas. Also, the reporting of financial support for non-Annex I parties is generally unsatisfactory, although it is improving. A dialogue continues among state parties to the FCCC about whether a compliance system with powers to penalize states for non-compliance with reporting and mitigation rules is a necessary element of accountable reporting. Some questions of practice and design surrounding accountable reporting in the climate change regime thus persist. It therefore seems premature to attempt to build a case for the rule's customary status.

While accountable reporting is only a procedural rule of the regime and does not grab headlines, the transparency of action it generates very likely keeps up a pressure on states to individualize the FCCC's general mitigation rule. By demanding accountability, it is potentially an effective weapon against free riding and mistrust. Moreover, effecting equitable burden-sharing of the mitigation responsibility among states, which is a requirement of the general mitigation rule, depends on the implementation of the accountable reporting rule, because transparent, accurate, and complete reporting is a precondition for calculations of comparability of effort. In sum, the accountable reporting rule has already achieved such normativity that it is almost inconceivable that it would be wound back.

Substantive climate law also offers some firm ground. Here, again, I summarize my findings at the risk of repetition. The mitigation of collective state emissions to a level at which dangerous climate change is prevented is a binding rule of climate change treaty law, as well as its ultimate objective. This prevention rule has existed since the FCCC went into force. Clearly, it is communal in form. Soon after the rule was instituted, state parties clarified 'dangerous' to mean warming of 2°C or over. The normative force behind the rule is clear: 'preventive' mitigation means desisting from the destruction of life as we know it, including human life and culture, through the deliberate alteration of the earth's climate. There could be no rule more deeply embedded in our values.

A good-faith obligation attaches to states individually, to the effect that they must strive to individualize the general mitigation rule and otherwise conduct themselves consistently with it. While this rule is well aligned with the type of problem it seeks to address, it creates a grey area of legal responsibility. This will persist for as long as the imperative to particularize state mitigation obligations through a burden-sharing agreement remains elusive. In its communal form, as it is expressed in the FCCC, the general mitigation rule may have already achieved the status of customary international law—which would be remarkable considering that the Convention came into effect only 20 years ago and the problem of climate change was little understood before then. Certainly, no state would dissent from the general mitigation rule. Mainstream international fora, including UN bodies, repeatedly reaffirm its status as law. However, the fact that states to date have failed to comply with the rule's central thrust (as explained below) would tend to undermine any claim to the rule's customary status.

The greater mitigation rule awaits elaboration by states for its full effect to be unleashed. The greatest challenge is to reach agreement on (a) a global budget of greenhouse gas emissions still permitted, (b) the period over which states are to expend it, (c) its specific allocation among states, and (d) the method by which to lock this obligation into place while ensuring that the budget and individual state allocations remain revisable in light of the evolving science of climate change. As we have seen, however, there has been no development of the general mitigation rule in the direction of a rule of law binding states to specific emission targets formulated with reference to an overall emission budget (top-down targets) calculated to avoid dangerous climate change, thus bringing about compliance with the prevention rule. The Kyoto Protocol did not employ top-down targets, and hence made no contribution to the further development of 'preventive' international climate law. As discussed in Chapter 4, the targets for both periods were domestically determined (bottom-up targets) and limited to those Annex I parties supporting the Protocol (a diminishing number). The FCCC, in the period

2013–2020, broadened state participation, but it too has employed ad hoc bottom-up targets and thus has not contributed to the individualization of the general mitigation rule.

'New and additional' financial support channelled from Annex I to non-Annex I parties to enable accountable reporting, reduction in emissions, and other regime outcomes in the latter group is a binding rule of climate change treaty law. It has the appearance of a procedural or facilitative rule (finance is input, not output). Yet, because its aim is mitigation through the delivery of finance, it is best thought of as an aspect of the prevention rule. Like the prevention rule, it takes a communal form. It requires a global budget for financial support, and it requires the budget to be informed by the additional mitigation effort that needs to occur in developing countries to avoid the 2°C threshold. It also requires a burden-sharing agreement that equitably assesses state-level financial contributions to the budget by Annex I parties, as well as a rule binding those parties to their assessed contributions and the recipient states to carrying out the funded actions.

As with particularization of the prevention rule, the financial-support rule also has not been developed beyond its communal form.

Having clarified the applicable legal rules, my argument proceeds to determine the extent of state compliance. Compliance with accountable reporting, as already noted, is high in the area of emission reporting, especially for Annex I parties. Here, the rule is individualized, fundamentally logical and necessary, and does not raise any equity issues—hence the widespread compliance. The FCCC parties have steadily improved the thoroughness of the applicable reporting rules. It has become standard practice to subject Annex I national communications and greenhouse gas inventories to review by Expert Review Teams. The reviews regularly identify areas of improvement. Rarely have they identified any significant or deliberate non-compliance. For the 2013–2020 period, the FCCC parties have introduced additional reporting on mitigation pledges, for both Annex I and non-Annex I parties. Non-Annex I reporting has always in a sense been obligatory, although only to the extent that circumstances allow. Often, the circumstances have not allowed for much. Non-Annex I reporting has been hampered by low capacity, competing developmental priorities, and insufficient funds. In this light, there is little evidence to be found of state non-compliance with reporting on emissions. There is as well a clear trend towards universal compliance with the rule, even if that level of compliance is a decade or more away. Despite the states' remarkable success in their application of the accountable reporting rule, there are still pockets of practice (viz. financial transfer, the CDM, and REDD) that are out of step with what accountable reporting demands. In these areas, specifically, states could said to be in non-compliance with the rule.

Accountable reporting under the FCCC is not subject to a formal compliance system. The Kyoto Protocol, by contrast, allows for states to be penalized for shortcomings in their reporting. This difference has raised a question about whether the Protocol's arrangement is superior to the FCCC's. Is accountable reporting incomplete in the absence of a strong compliance response? I have argued that the available evidence does not answer the question one way or the other. To start with, the Protocol parties have not 'complied' very well with their own compliance system. They have allowed the Facilitative Branch to fall into dysfunction. As for the Enforcement Branch, it has engaged with relatively minor issues that could just as well have been corrected by the states themselves on advice coming directly from the Expert Review Teams. The highest penalty at the Enforcement Branch's disposal, which relates to the mitigation obligation rather than to reporting, has not been a real threat to Annex I parties. This is primarily because the Protocol's targets are bottom-up and unambitious, but it is also because the targets are unconstrained by any objective reference point. A state penalized for exceeding its target in one commitment period can get around the penalty by volunteering a less ambitious target for the subsequent commitment period.

In sum, the Protocol's 'compliance-enhanced' accountable reporting has not, on the evidence, demonstrated an advantage over the FCCC's version.

There have been two completed periods during which individualized mitigation rules (and not just the FCCC's general mitigation rule) have applied to Annex I states. The first period (1990–2000, or, more accurately, 2000 emissions compared with 1990 emissions) was under the control of the FCCC's specific mitigation rule. The second period (2008–2012) was controlled by the Protocol's first commitment period.

The FCCC's specific mitigation rule was complied with only if EIT emissions are included in the total. If the great drop in emissions in EIT countries is taken out of the total, the emissions of the remaining Annex I parties in 2000 were higher than they were in 1990, breaching the rule. Still, the rule allowed for the target to be met 'individually or jointly', so in a literal sense there was no breach of the specific mitigation rule.

For the Protocol's 2008–2012 period there was compliance with the period's mitigation targets by a large majority of Annex I parties even before any use was made of the flexibility mechanisms to compensate for shortfalls. An exception worth emphasizing is Canada's performance. The country failed by a large margin to comply with its target. It withdrew from the treaty, perhaps to avoid a formal finding of non-compliance, but more likely because its government by that point had become fully opposed to the Protocol and had sought to align its position with that of the United States. A further qualification of the success of this period is that the Protocol targets of EIT countries did not present any practical limitation to emission growth in those countries, and no action was taken to adjust the targets to reflect the new emission profile of EITs. They thus 'complied' effortlessly.

Although the Protocol's first commitment period represents the high-water mark of state achievement on cooperative mitigation, the evidence of the effort is too shallow to support a generalization about what states are capable of, or inclined to do, under such arrangements. The fact that many Annex I parties came to reject the Protocol model even as they continued to comply with their obligations under it could be read positively to mean that compliance would be even higher under a model they supported.

As for compliance with the FCCC's general mitigation rule, the rule's several aspects need to be taken in turn. In a trivial sense, the core of the rule has not been breached because global warming has not yet exceeded the 2°C threshold. The opposite conclusion is reached through a more realistic analysis of what the rule demands. States collectively must direct their emissions onto a trajectory that is consistent with a stabilization of temperature rise at a level below the 2°C limit. The number of realistically achievable pathways diminishes year by year. If states are not currently on a qualifying trajectory, they are, on this analysis, in breach of the general mitigation rule. The 'emission gap' for 2020 (the difference between emission levels in 2020 consistent with meeting the 2°C target and emission levels in the same year if countries do not deepen their current mitigation pledges) is the most salient measure of what states are *not* doing to comply in substance. States collectively are therefore in non-compliance with a central element of the general mitigation rule.

I have also pointed out that the FCCC's communal rule entails an obligation that falls on states individually to work faithfully to particularize the rule. Here, the compliance of states has many sides to it. The FCCC parties continue to negotiate, and no party has left the fold of the process. But this is the most that could be said on the positive side. On the negative side, many parties seem to regard the 2013–2020 period as closed to further negotiation on mitigation 'ambition'. I include in this category both Annex I parties (who should be increasing their ambition) and non-Annex I parties (who should be making pledges to bring down business-as-usual emissions in a measurable way). I have argued that the 2013–2020 pledges under the FCCC, which belong mainly to Annex I parties but are not limited to them, are best conceptualized as a halfway house in the implementation of the general mitigation rule. The element that must be added for there to be progress is for each state to accept a top-down target representing a slice of a global emission budget. There is no indication that such an element will be added to an agreement anytime soon. Moreover, there is hardly any evidence of progress being made on a formula for equitable burden-sharing among states.

Looking to the future, it is, of course, premature to speculate about whether states will comply with their pledged 2013–2020 targets. Going on past evidence, most Annex I states will probably comply. Because the targets are ad hoc (bottom-up) and too vague in the case of many of the non-Annex I parties, compliance with the prevention rule will not be advanced in this period irrespective of the extent of compliance.

In the matter of compliance enforcement for mitigation, the systems that have been put in place to promote state compliance under both the FCCC and the Kyoto Protocol have not been relevant to enforcing the prevention rule (or the financial-support rule), in the sense of policing adherence to top-down targets. This is because top-down targets, as already noted, have never been developed under either treaty. States are a long way from developing a compliance system for the general mitigation rule. They might never develop one, if the view prevails that a 'strict' compliance system provides no net benefit.

When discussing compliance with reporting obligations I referred to my conclusion that the model of binding bottom-up targets coupled with penalties for non-compliance which has been the Kyoto Protocol's model has not been an unequivocal success. (Canada overshot its target despite the compliance system. Other countries also did so, to a lesser extent; a glut of cheap allowances will enable them easily to make up the difference.) Because the Protocol's compliance system has not demonstrated an added value, it is being eclipsed by the view that accountable reporting, bolstered at most by an IAR/ICA-type level of scrutiny of state reports, is sufficient to secure compliance with emission targets, especially with bottom-up targets that by definition have been self-determined by states.

The view just described might not seem contentious when confined to bottom-up pledges. But what if tougher targets were involved, i.e. targets derived from an emission budget that aims to close the 2020 emission gap? Would accountable reporting without a separate compliance system still be sufficient? Because we have had no experience with this situation, we have no evidence one way or the other. Several scholars have argued that effective monitoring and enforcement are required to counteract state incentives to avoid treaty obligations.[1] A contrary view, discussed in Chapter 5, is that of the US government, which maintains that state reporting and international discussion of state reports are sufficient as a compliance system within the climate change regime. According to this view, enforcement and formal sanctions are not required for compliance, not because states are not obliged to deliver on their commitments but because accountable reporting is a better approach to compliance in cases where it is also a requirement that *all* states take on legally binding commitments. It might be that the critical factor finally is not the extent of the emission-reduction burden but the degree of fairness with which it has been allocated to each state. Where an 'individualization' agreement is universally perceived as fair (which is not true of the Protocol), the FCCC's version of accountable reporting (plus a form of IAR/ICA) might be sufficient for compliance, even though (or perhaps precisely because) the targets are not self-selected but have been determined by the application of a fair allocation formula to a global emission budget. The theory here would be that states would try harder to do their part under an equitable agreement, knowing that if they fall short they cannot blame the agreement. If trading is enabled in this system, a state that falls short would prefer to buy itself out of trouble than be accused of unreliability or lack of integrity. Canada might not have abandoned its target had the target been rationally worked out and had it constituted a contribution to a meaningful outcome.

---

1 For example, George W. Downs (1998), 'Enforcement and the Evolution of Cooperation', 19 *Michigan Journal of International Law* 319. According to Hare and colleagues, deep commitments require a strong compliance regime: William Hare et al. (2010), 'The Architecture of the Global Climate Regime: A Top-Down Perspective', 10 (6) *Climate Policy* 600, p. 608. However, if states could make strong emission-reduction commitments that are sufficient for the 2°C threshold and shared the burden equitably, a compliance system might be unnecessary, as long as there is access to reliable information about implementation (i.e. accountable reporting). This point is further explored in the main text, below.

In Chapter 3 we saw that, as of 2014, considerable doubt existed that a post-2020 agreement would allow for a Protocol-type compliance system. Moreover, it does not seem likely that top-down target setting will be pursued in the new agreement, at least not in the early phase of it. If all this is true, we are still a long way from testing the theory that a just agreement will work well without a strict compliance system.

Two opt-in treaty mechanisms were considered in this book: the CDM under the Kyoto Protocol and REDD under the FCCC. A majority of states have participated in the CDM; as for REDD, it is likely that all Annex I parties will eventually become bound to contribute financially to the programme. Both schemes promise mitigation. Hence, compliance with the 'optional' systems is a question of general concern.

Two rules of treaty law affect the CDM. I have argued that compliance with the additionality rule is, at best, difficult to assess because of the high degree of speculation inherent in the determination of additionality. Compliance with the CDM's second binding element, the sustainable development rule, is also difficult to assess. It appears to have been ignored by states as being too difficult to conceptualize or account for. These problems are not simply matters of non-compliance with technical rules of the CDM. They also imply non-compliance with the accountable reporting and general mitigation rules. In this specialized domain of the Kyoto Protocol, states has taken a casual approach to reporting and mitigation. It seems to be enough for them that the CDM is working and producing offsets.

It is too early to talk about compliance within the framework of the REDD system. My discussion on this point has, therefore, been pre-emptive. REDD may be suffering from the same 'some action is better than no action' affliction of the CDM. The climate change regime is eager to get projects up and running quickly—or, in the case of REDD, tested on a small scale and progressively scaled up. The increasing urgency for action on the ground has supported a tendency to hurry schemes out before accountable reporting systems are decided or developed. For the most part, REDD is destined for contexts of weak government and poor transparency. The FCCC parties are proceeding with REDD even though the evidence suggests that it will not produce reliable mitigation outcomes in the short term. Moreover, they are proceeding with the scheme even though the sources of funding are still undecided. My discussion in Chapter 6 supports the conclusion that states, by pursuing REDD prematurely, risk further violations of the accountable reporting and prevention rules. Even if REDD were kept as a non-market system (i.e. was not allowed to produce offsets), it will still be difficult to tell how much mitigation it delivers and whether the money is well spent from a mitigation point of view (compared with alternative mitigation strategies).

To recapitulate, the climate change regime can claim a foundation in two binding, normatively strong, rules of law. Compliance is high in matters of accountable reporting but low in relation to the prevention rule and its included financial-support rule. The cause of non-compliance with the prevention rule is that there has been no serious discussion—let alone an agreement—on effort-sharing of top-down targets referenced to a 2°C emission budget. Absent a common position on a mitigation pathway or on how emission reductions should be shared among states to realize that pathway for the collective of states, individual states continue to pursue unambitious mitigation practices (and financial transfers), while the negotiations drag on and the remaining options become fewer and harder for all.

Why are states unable to advance the general mitigation rule? The causes are certainly not legal ones. Strictly, this means that they lie beyond the scope of this book. However, they are not so abstruse as to prevent a brief comment. First, the communal form of the general mitigation rule means that states can use the inaction of others as an excuse for their own inaction. Second, there is a clear lack of bottom-up pressure on governments to act on climate change. As *Nature* has put it, 'What is missing from [the current] talks is not science but political ambition, which is ultimately a reflection of public support.'[2] Climate change would have to be widely regarded as a high-priority problem in order that personal sacrifice in the present is palatable. Yet, majorities of voters in most countries perceive climate change as mostly a natural phenomenon and beyond human control.[3] Increasingly dramatic changes in the Arctic, but also in the Antarctic and elsewhere, have drawn muted responses from the public. Ambitious policy on climate change means a policy that substantially raises consumer prices.[4] To this, the response is not muted. Most citizens are now well aware of the link. Concern about rising consumer prices and a commitment to rapid economic development remain high, and these priorities are not necessarily compatible with strong action on climate change. The past provides evidence that people will choose to adjust to nature's new extremes rather than choose more costly or limiting ways of life. The climate change regime hopes to manage forces (natural, but also human) that might finally prove too powerful to control.[5] Even the 2°C plan is hubristic. It presumes that nature can safely be modified for the sake of preserving a presently preferred way of living. It is a high-risk gamble. New normative law—in particular the individualization of the general mitigation rule—will not easily develop in a context where regulation is thought to be excessively costly, and it will be even harder to realize if the phenomenon to be regulated continues to be doubted. We must conclude, I think, that international climate law is likely to remain static for many years to come.

---

2 Editorial (19 September 2013), 'The Final Assessment', 501 *Nature* 281, p. 281. See also Elliot Diringer (19 September 2013), 'A Patchwork of Emissions Cuts', 501 *Nature* 307, p. 308.

3 Linda H. Connor and Nick Higginbotham (2013), '"Natural Cycles" in Lay Understandings of Climate Change', 23 *Global Environmental Change* 1852, p. 1858.

4 For the general rule, see Susanne Olbrisch et al. (2011), 'Estimates of Incremental Investment for and Cost of Mitigation Measures in Developing Countries', 11 (3) *Climate Policy* 970, e.g. p. 973 ('Low-carbon generation technologies are relatively capital-intensive and the incremental capital cost is only partially offset by lower operating costs, leading to higher electricity prices').

5 For a parallel case on a smaller scale, see the chapter 'Atchafalaya', in John McPhee (1989), *The Control of Nature*.

Bibliography

1. Scholarly works

Abbs, Ross, Peter Cashman, and Tim Stephens (2012). 'Australia.' In *Climate Change Liability: Transnational Law and Practice*, edited by Richard Lord, Silke Goldberg, Lavanya Rajamani, and Jutta Brunnée. Cambridge, UK: Cambridge University Press, 67–111.

Aguilar, Soledad, and Eugenia Recio (2013). 'Climate Law in Latin American Countries.' In *Climate Change and the Law*, edited by Erkki J. Hollo, Kati Kulovesi, and Michael Mehling. Dordrecht: Springer, 653–678.

Alexeew, Johannes, Linda Bergset, Kristin Meyer, Juliane Petersen, Lambert Schneider, and Charlotte Unger (2010). 'An Analysis of the Relationship between the Additionality of CDM Projects and their Contribution to Sustainable Development.' 10 *International Environmental Agreements* 233–248.

Allen, Myles, Pardeep Pall, Daithi Stone, Peter Stott, David Frame, Seung-Ki Min, Toru Nozawa, and Seiji Yukimoto (2007). 'Scientific Challenges in the Attribution of Harm to Human Influence on Climate.' 155(6) *University of Pennsylvania Law Review* 1353–1400.

Andersson, Krister, Tom P. Evans, and Kenneth R. Richards (2009). 'National Forest Carbon Inventories: Policy Needs and Assessment Capacity.' 93 *Climatic Change* 69–101.

Andresen, Steinar (2014). 'The Climate Regime: A Few Achievements, but Many Challenges.' 4(1–2) *Climate Law* 21–29.

Andresen, Steinar, and Lars H. Gulbrandsen (2005). 'The Role of Green NGOs in Promoting Climate Compliance.' In *Implementing the Climate Regime: International Compliance*, edited by Jon Hovi, Olav Stokke, and Geir Ulfstein. London: Earthscan, 169–186.

Bakker, Stefan, Constanze Haug, Harro Van Asselt, Joyeeta Gupta, and Raouf Saïdi (2011). 'The Future of the CDM: Same Same, But Differentiated?' 11 *Climate Policy* 752–767.

Barrett, Scott (2008). 'Climate Treaties and the Imperative of Enforcement.' 24(2) *Oxford Review of Economic Policy* 239–258.

Barratt-Brown, Elizabeth P. (1991). 'Building a Monitoring and Compliance Regime Under the Montreal Protocol.' 16(2) *Yale Journal of International Law* 519–570.

Bellassen, Valentin, and Sebastiaan Luyssaert (2014). 'Managing Forests in Uncertain Times.' 506 *Nature* 153–155.

Berntsen, Terje, Jan Fuglestvedt, and Frode Stordal (2005). 'Reporting and Verification of Emissions and Removals of Greenhouse Gases.' In *Implementing the Climate Regime:*

International Compliance, edited by Jon Hovi, Olav Stokke, and Geir Ulfstein. London: Earthscan, 85–105.

Beyerlin, Ulrich (2007). 'Different Types of Norms in International Environmental Law: Policies, Principles, and Rules.' In *The Oxford Handbook of International Environmental Law*, edited by Daniel Bodansky, Jutta Brunnée, and Ellen Hey. Oxford: Oxford University Press, 425–448.

Bietta, Federica (2010). 'From the Hague to Copenhagen: Why it Failed then and Why it Could be Different.' In *Deforestation and Climate Change: Reducing Carbon Emissions from Deforestation and Forest Degradation*, edited by Valentina Bosetti and Ruben Lubowski. Cheltenham, UK: Edward Elgar, 27–38.

Bilder, Richard B. (2000). 'Beyond Compliance: Helping Nations Cooperate.' In *Commitment and Compliance: The Role of Non-Binding Norms in the International Legal System*, edited by Dinah Shelton. Oxford: Oxford University Press, 65–73.

Bodansky, Daniel (2010). *The Art and Craft of International Environmental Law.* Cambridge, MA: Harvard University Press.

Boyd, Emily, Nate Hultman, J. Timmons Roberts, Esteve Corbera, John Cole, Alex Bozmoski, Johannes Ebeling, Robert Tippman, Philip Manna, Katrina Brown, and Diana M. Liverman (2009). 'Reforming the CDM for Sustainable Development: Lessons Learned and Policy Futures.' 12(7) *Environmental Science and Policy* 820–831.

Bradford, William (2005). 'International Legal Compliance: Surveying the Field.' 36(2) *Georgetown Journal of International Law* 495–536.

Brown, Chester (2010). 'International, Mixed, and Private Disputes Arising under the Kyoto Protocol'. 1(2) *Journal of International Dispute Settlement* 447–473.

Brunnée, Jutta (2000). 'A Fine Balance: Facilitation and Enforcement in the Design of a Compliance Regime for the Kyoto Protocol.' 13(2) *Tulane Environmental Law Journal* 223–270.

Brunnée, Jutta (2003). 'The Kyoto Protocol: Testing Ground for Compliance Theories?' 63 *Zeitschrift für Ausländisches Öffentliches Recht und Völkerrecht* 255–280.

Brunnée, Jutta (2012a). 'Climate Change and Compliance and Enforcement Processes.' In *International Law in the Era of Climate Change*, edited by Rosemary Rayfuse and Shirley V. Scott. Cheltenham, UK: Edward Elgar, 290–320.

Brunnée, Jutta (2012b). 'Promoting Compliance with Multilateral Environmental Agreements.' In *Promoting Compliance in an Evolving Climate Regime*, edited by Jutta Brunnée, Meinhard Doelle, and Lavanya Rajamani. Cambridge, UK: Cambridge University Press, 38–54.

Brunnée, Jutta, Silke Goldberg, Richard Lord, and Lavanya Rajamani (2012). 'Overview of Legal Issues Relevant to Climate Change.' In *Climate Change Liability: Transnational Law and Practice*, edited by Richard Lord, Silke Goldberg, Lavanya Rajamani, and Jutta Brunnée. Cambridge, UK: Cambridge University Press, 23–49.

Brunnée, Jutta, and Stephen J. Toope (2010). *Legitimacy and Legality in International Law: An Interactional Account.* Cambridge, UK: Cambridge University Press.

Brunnée, Jutta, and Stephen J. Toope (2011). 'Interactional International Law: An Introduction.' 3(2) *International Theory* 307–318.

Burgstaller, Markus (2005). *Theories of Compliance with International Law.* Leiden: Martinus Nijhoff.

Burns, William C.G., and Hari M. Osofsky (2009). 'The Exigencies that Drive Potential Causes of Action for Climate Change.' In *Adjudicating Climate Change: State, National, and International Approaches*, edited by William C.G. Burns and Hari M. Osofsky. New York: Cambridge University Press, 1–27.

Caldeira, Ken, and Steven J. Davis (2011). 'Accounting for Carbon Dioxide Emissions: A Matter of Time.' 108(21) *PNAS* 8533–8534.

Campbell, David (2013). 'After Doha: What Has Climate Change Policy Accomplished?' 25(1) *Journal of Environmental Law* 125–136.

Carlarne, Cinnamon (2012). 'The Future of the UNFCCC: Adaptation and Institutional Rebirth for the International Climate Convention.' Working Paper Series No. 172, Center for Interdisciplinary Law and Policy Studies, Moritz College of Law, 1–58.

Chayes, Abram, and Antonia Handler Chayes (1995). *The New Sovereignty: Compliance with International Regulatory Agreements*. Cambridge, MA: Harvard University Press.

Chayes, Abram, Antonia Handler Chayes, and Ronald B. Mitchell (1998). 'Managing Compliance: A Comparative Perspective.' In *Engaging Countries: Strengthening Compliance with International Environmental Accords*, edited by Edith Brown Weiss and Harold K. Jacobson. Cambridge, MA: MIT Press, 39–62.

Chayes, Antonia Handler, Abram Chayes, and Ronald B. Mitchell (1995). 'Active Compliance Management in Environmental Treaties.' In *Sustainable Development and International Law*, edited by Winfried Lang. London: Graham and Trotman, 75–89.

Checkel, Jeffrey T. (2001). 'Why Comply? Social Learning and European Identity Change.' 53(3) *International Organization* 553–588.

Churchill, Robin R., and Geir Ulfstein (2000). 'Autonomous Institutional Arrangements in Multilateral Environmental Agreements: A Little-Noticed Phenomenon in International Law.' 94 *American Journal of International Law* 623–659.

Ciais, P., P. Rayner, F. Chevallier, P. Bousquet, M. Logan, P. Peylin, and M. Ramonet (2010). 'Atmospheric Inversions for Estimating CO_2 Fluxes: Methods and Perspectives.' 103 *Climatic Change* 69–92.

Connor, Linda H., and Nick Higginbotham (2013). '"Natural cycles" in Lay Understandings of Climate Change.' 23 *Global Environmental Change* 1852–1861.

Cox, Prentiss (2011). *Analysis of Cookstove Change-Out Projects Seeking Carbon Credits*. Environmental Sustainability Clinic, University of Minnesota Law School.

Cummings, Brendan R., and Kassie R. Siegel (2009). 'Biodiversity, Global Warming, and the United States Endangered Species Act: The Role of Domestic Wildlife Law in Addressing Greenhouse Gas Emissions.' In *Adjudicating Climate Change: State, National, and International Approaches*, edited by William C.G. Burns and Hari M. Osofsky. New York: Cambridge University Press, 145–172.

Dai, Xinyuan (2005). 'Why Comply? The Domestic Constituency Mechanism.' 59 *International Organization* 363–398.

Danish, Kyle (1997). 'Review of A. Chayes and A.H. Chayes, *The New Sovereignty*.' 37 *Virginia Journal of International Law* 789–810.

Dannenmaier, Eric (2012). 'The Role of Non-State Actors in Climate Compliance.' In *Promoting Compliance in an Evolving Climate Regime*, edited by Jutta Brunnée, Meinhard Doelle, and Lavanya Rajamani. Cambridge, UK: Cambridge University Press, 149–176.

De Cendra de Larragán, Javier (2011). 'Liability of Member States and the EU in View of the International Climate Change Framework: Between Solidarity and Responsibility.' In *Climate Change Liability*, edited by Michael Faure and Marjan Peeters. Cheltenham, UK: Edward Elgar, 55–89.

De Sadeleer, Nicolas (2002). *Environmental Principles: From Political Slogans to Legal Rules*. Oxford: Oxford University Press.

Diringer, Elliot (2013). 'A Patchwork of Emissions Cuts.' 501 *Nature* 307–309.

Disch, David (2010). 'A Comparative Analysis of the "Development Dividend" of Clean Development Mechanism Projects in Six Host Countries.' 2 *Climate and Development* 50–64.

Doelle, Meinhard (2004). 'The Kyoto Protocol: Reflections on its Significance on the Occasion of its Entry into Force.' 27(2) *Dalhousie Law Journal* 555–566.

Doelle, Meinhard (2005). *From Hot Air to Action? Climate Change, Compliance and the Future of International Environmental Law.* Thomson Carswell.

Doelle, Meinhard (2010). 'Early Experience with the Kyoto Compliance System: Possible Lessons for MEA Compliance System Design.' 1(2) *Climate Law* 237–260.

Doelle, Meinhard (2013). 'Compliance and Enforcement in the Climate Change Regime.' In *Climate Change and the Law*, edited by Erkki J. Hollo, Kati Kulovesi, and Michael Mehling. Dordrecht: Springer, 165–188.

Doelle, Meinhard, Dennis Mahony, and Alex Smith (2012). 'Canada.' In *Climate Change Liability: Transnational Law and Practice*, edited by Richard Lord, Silke Goldberg, Lavanya Rajamani, and Jutta Brunnée. Cambridge, UK: Cambridge University Press, 525–555.

Downs, George W. (1998). 'Enforcement and the Evolution of Cooperation.' 19 *Michigan Journal of International Law* 319–344.

Drumbl, Mark A. (2002). 'Poverty, Wealth, and Obligation in International Environmental Law.' 76(4) *Tulane Law Review* 843–960.

Dubash, Navroz K., Markus Hagemann, Niklas Höhne, and Prabhat Upadhyaya (2013). 'Developments in National Climate Change Mitigation Legislation and Strategy.' 13 *Climate Policy* 649–664.

Durant, A.J., Corinne Le Quéré, C. Hope, and A.D. Friend (2011). 'Economic Value of Improved Quantification in Global Sources and Sinks of Carbon Dioxide.' 369 *Philosophical Transactions of the Royal Society A* 1967–1979.

Ehrmann, Markus (2002). 'Procedures of Compliance Control in International Environmental Treaties'. 13(2) *Colorado Journal of International Environmental Law and Policy* 377–443.

Ekardt, Felix (2013). 'Climate Law in Germany.' In *Climate Change and the Law*, edited by Erkki J. Hollo, Kati Kulovesi, and Michael Mehling. Dordrecht: Springer, 523–536.

Ellis, Jane, Sara Moarif, and Greg Briner (2010). *Core Elements of National Reports*. OECD/International Energy Agency.

Ellis, Jane, Harald Winkler, Jan Corfee-Morlot, and Frédéric Gagnon-Lebrun (2007). 'CDM: Taking Stock and Looking Forward.' 35(1) *Energy Policy* 15–28.

Farber, Daniel A. (2008). 'The Case for Climate Compensation: Justice for Climate Change Victims in a Complex World.' 2008(2) *Utah Law Review* 377–413.

Faure, Michael G. (2012). 'Effectiveness of Environmental Law: What Does the Evidence Tell Us?' 36 *William and Mary Environmental Law and Policy Review* 293–336.

Faure, Michael G., and Jürgen Lefevere (2011). 'Compliance with Global Environmental Policy.' In *The Global Environment: Institutions, Law, and Policy* (3rd edn), edited by Regina S. Axelrod, Stacy D. VanDeveer, and David Leonard Downie. Washington DC: CQ Press, 172–191.

Finnemore, Martha (2000). 'Are Legal Norms Distinctive?' 32 *New York University Journal of International Law and Politics* 699–705.

Finus, Michael (2008). 'The Enforcement Mechanisms of the Kyoto Protocol: Flawed or Promising Concepts?' 1 *Letters in Spatial and Resource Sciences* 13–25.

Fitzmaurice, Malgosia, and Catherine Redgwell (2000). 'Environmental Non-Compliance Procedures and International Law.' 31 *Netherlands Yearbook of International Law* 35–65.

Franck, Thomas M. (1995). *Fairness in International Law and Institutions*. Oxford: Clarendon Press.
Fransen, Taryn (2009). *Enhancing Today's MRV Framework to Meet Tomorrow's Needs: The Role of National Communications and Inventories*. Washington DC: World Resources Institute.
Fraser, Barbara (2014). 'Carving Up the Amazon.' 509 *Nature* 418–419.
French, Duncan, and Lavanya Rajamani (2013). 'Climate Change and International Environmental Law: Musings on a Journey to Somewhere.' 25(3) *Journal of Environmental Law* 437–461.
Gerardu, Jo, Meredith Koparova, Ken Markowitz, Elise Stull, and Durwood Zaelke (eds) (2013). *Compliance Strategies to Deliver Climate Benefits*. Washington DC: Institute for Governance and Sustainable Development.
Gerrard, Michael B., and Gregory E. Wannier (2012). 'United States of America.' In *Climate Change Liability: Transnational Law and Practice*, edited by Richard Lord, Silke Goldberg, Lavanya Rajamani, and Jutta Brunnée. Cambridge, UK: Cambridge University Press, 556–603.
Giesberts, Ludger, Alexander Sarac, and John Wunderlin (2011). 'The Institutional Design of the CDM Appeals Body: Recent Developments and Key Considerations.' 5(2) *Carbon and Climate Law Review* 277–286.
Gillenwater, Michael, and Stephen Seres (2011). *The Clean Development Mechanism: A Review of the First International Offset Program*. Arlington, VA: Pew Center on Global Climate Change.
Glennon, Michael J., and Alison L. Stewart (1998). 'The United States: Taking Environmental Treaties Seriously.' In *Engaging Countries: Strengthening Compliance with International Environmental Accords*, edited by Edith Brown Weiss and Harold K. Jacobson. Cambridge, MA: MIT Press, 173–213.
Goldberg, Silke, and Richard Lord (2012). 'England.' In *Climate Change Liability: Transnational Law and Practice*, edited by Richard Lord, Silke Goldberg, Lavanya Rajamani, and Jutta Brunnée. Cambridge, UK: Cambridge University Press, 445–488.
Goldsmith, Arthur A. (2007). 'Is Governance Reform a Catalyst for Development?' 20(2) *Governance* 165–186.
Goldsmith, Jack L., and Eric A. Posner (1999). 'A Theory of Customary International Law.' 66(4) *University of Chicago Law Review* 1113–1177.
Goldsmith, Jack L., and Eric A. Posner (2003). 'International Agreements: A Rational Choice Approach.' 44(1) *Virginia Journal of International Law* 113–143.
Gouritin, Armelle (2011). 'Potential Liability of European States Under the ECHR for Failure to Take Appropriate Measures with a View to Adaptation to Climate Change.' In *Climate Change Liability*, edited by Michael Faure and Marjan Peeters. Cheltenham, UK: Edward Elgar, 134–162.
Grasso, Marco, and J. Timmons Roberts (2014). 'A Compromise to Break the Climate Impasse.' 4 *Nature Climate Change* 543–549.
Grindle, Merilee S. (2004). 'Good Enough Governance: Poverty Reduction and Reform in Developing Countries.' 17(4) *Governance* 525–548.
Gusti, Mykola, and Matthias Jonas (2010). 'Terrestrial Full Carbon Account for Russia: Revised Uncertainty Estimates and their Role in a Bottom-up/Top-Down Accounting Exercise.' 103 *Climatic Change* 159–174.
Guzman, Andrew T. (2002). 'A Compliance Based Theory of International Law.' 90(6) *California Law Review* 1823–1887.

Haas, Peter M. (2000). 'Choosing to Comply: Theorizing from International Relations and Comparative Politics.' In *Commitment and Compliance: The Role of Non-Binding Norms in the International Legal System*, edited by Dinah Shelton. Oxford: Oxford University Press, 43–64.

Haites, Erik, and Farhana Yamin (2004). 'Overview of the Kyoto Mechanisms.' 5(1) *International Review for Environmental Strategies* 199–261.

Halvorssen, Anita, and Jon Hovi (2006). 'The Nature, Origin and Impact of Legally Binding Consequences: The Case of the Climate Regime.' 6 *International Environmental Agreements* 157–171.

Handl, Günter (1997). 'Compliance Control Mechanisms and International Environmental Obligations.' 5 *Tulane Journal of International and Comparative Law* 29–49.

Hare, William, Claire Stockwell, Christian Flachsland, and Sebastian Oberthür (2010). 'The Architecture of the Global Climate Regime: A Top-Down Perspective.' 10(6) *Climate Policy* 600–614.

Haritz, Miriam (2011). 'Liability *With* and Liability *From* the Precautionary Principle in Climate Change Cases.' In *Climate Change Liability*, edited by Michael Faure and Marjan Peeters. Cheltenham, UK: Edward Elgar, 15–46.

Harris, Paul G. (2014). 'Risk-Averse Governments.' 4 *Nature Climate Change* 245–246.

Hathaway, Oona A. (2005). 'Between Power and Principle: An Integrated Theory of International Law.' 72(2) *University of Chicago Law Review* 469–536.

Herold, Anke (2012). 'Experiences with Articles 5, 7, and 8: Defining the Monitoring, Reporting and Verification System Under the Kyoto Protocol.' In *Promoting Compliance in an Evolving Climate Regime*, edited by Jutta Brunnée, Meinhard Doelle, and Lavanya Rajamani. Cambridge, UK: Cambridge University Press, 122–146.

Hilson, Chris (2013). 'It's All About Climate Change, Stupid! Exploring the Relationship Between Environmental Law and Climate Law.' 25(3) *Journal of Environmental Law* 359–370.

Horstmann, Britta, and Achala Chandani Abeysinghe (2011). 'The Adaptation Fund of the Kyoto Protocol: A Model for Financing Adaptation to Climate Change?' 2(3) *Climate Law* 415–437.

Hovi, Jon, Camilla Bretteville Froyn, and Guri Bang (2007). 'Enforcing the Kyoto Protocol: Can Punitive Consequences Restore Compliance?' 33 *Review of International Studies* 435–449.

Hovi, Jon, Mads Greaker, Cathrine Hagem, and Bjart Holtsmark (2012). 'A Credible Compliance Enforcement System for the Climate Regime.' 12(6) *Climate Policy* 741–754.

Howlett, Michael (2014). 'Why are Policy Innovations Rare and so Often Negative? Blame-Avoidance and Problem-Denial in Climate Change Policy-Making.' *Global Environmental Change*, in press.

Huang, Peter H. (2002). 'International Environmental Law and Emotional Rational Choice.' 31(1) *Journal of Legal Studies* S237–S258.

Humphreys, Stephen (2012). 'Climate Change and International Human Rights Law.' In *International Law in the Era of Climate Change*, edited by Rosemary Rayfuse and Shirley V. Scott. Cheltenham, UK: Edward Elgar, 29–57.

Hunter, David B. (2009). 'The Implications of Climate Change Litigation: Litigation for International Environmental Law-Making.' In *Adjudicating Climate Change: State, National, and International Approaches*, edited by William C.G. Burns and Hari M. Osofsky. New York: Cambridge University Press, 357–374.

Isenberg, Jordan, and Catherine Potvin (2010). 'Financing REDD in Developing Countries: A Supply and Demand Analysis.' 10(2) *Climate Policy* 216–231.

Kalas, Peggy Rodgers, and Alexia Herwig (2000). 'Dispute Resolution under the Kyoto Protocol.' 27(1) *Ecology Law Quarterly* 53–133.

Kallbekken, Steffen, Håkon Sælen, and Arild Underdal (2014). *Equity and Spectrum of the Mitigation Commitments in the 2015 Agreement.* Copenhagen: Nordic Council of Ministers.

Kaminskaitė-Salters, Giedrė (2011). 'Climate Change Litigation in the UK: Its Feasibility and Prospects.' In *Climate Change Liability*, edited by Michael Faure and Marjan Peeters. Cheltenham, UK: Edward Elgar, 165–188.

Kaufmann, Daniel, Aart Kraay, and Pablo Zoido-Lobatón (2000). 'Governance Matters: From Measurement to Action.' 37(2) *Finance and Development* 10–13.

Keohane, Robert O. (1992). 'Compliance with International Commitments: Politics Within a Framework of Law.' 86 *American Society of International Law Proceedings* 176–179.

Keohane, Robert O., and Joseph S. Nye (1977). *Power and Interdependence: World Politics in Transition.* Boston, MA: Little, Brown.

Kerr, Richard A. (2014). 'Atlantic Current Can Shut Down For Centuries, Disrupting Climate.' 343 *Science* 831.

Kidd, Michael, and Ed Couzens (2013). 'Climate Change Responses in South Africa.' In *Climate Change and the Law*, edited by Erkki J. Hollo, Kati Kulovesi, and Michael Mehling. Dordrecht: Springer, 619–638.

Kim, Joy Aeree, Jane Ellis, and Sara Moarif (2009). *Matching Mitigation Actions with Support: Key Issues for Channelling International Public Finance.* OECD/International Energy Agency.

Kimura, Hitomi (2013). 'Climate Law and Policy in Japan.' In *Climate Change and the Law*, edited by Erkki J. Hollo, Kati Kulovesi, and Michael Mehling. Dordrecht: Springer, 585–595.

Kingsbury, Benedict (1998). 'The Concept of Compliance as a Function of Competing Conceptions of International Law.' 19 *Michigan Journal of International Law* 345–372.

Kishor, Nalin, and Arati Belle (2004). 'Does Improved Governance Contribute to Sustainable Forest Management?' 19(1–3) *Journal of Sustainable Forestry* 55–79.

Klabbers, Jan (2007). 'Compliance Procedures.' In *The Oxford Handbook of International Environmental Law*, edited by Daniel Bodansky, Jutta Brunnée, and Ellen Hey. Oxford: Oxford University Press, 995–1009.

Koh, Harold Hongju (1997). 'Why Do Nations Obey International Law?' 106 *Yale Law Journal* 2599–2659.

Koskenniemi, Martti (1990). 'The Pull of the Mainstream.' 88(6) *Michigan Law Review* 1946–1962.

Koskenniemi, Martti (1992). 'Breach of Treaty or Non-Compliance? Reflections on the Enforcement of the Montreal Protocol.' 3 *Yearbook of International Environmental Law* 123–162.

Koskenniemi, Martti (2002). *The Gentle Civilizer of Nations: The Rise and Fall of International Law 1870–1960.* Cambridge, UK: Cambridge University Press.

Krämer, Ludwig (2012). 'European Union Law.' In *Climate Change Liability: Transnational Law and Practice*, edited by Richard Lord, Silke Goldberg, Lavanya Rajamani, and Jutta Brunnée. Cambridge, UK: Cambridge University Press, 351–375.

Kravchenko, Svitlana (2007). 'The Aarhus Convention and Innovations in Compliance with Multilateral Environmental Agreements'. 18(1) *Colorado Journal of International Environmental Law and Policy* 1–50.

Kravchenko, Svitlana (2011). 'Giving the Public a Voice in MEA Compliance Mechanisms.' In *Compliance and Enforcement in Environmental Law: Toward More Effective Implementation*, edited by LeRoy Paddock, Du Qun, Louis J. Kotzé, David L. Markell, Kenneth J. Markowitz, and Durwood Zaelke. Cheltenham, UK: Edward Elgar, 83–110.

Krey, Matthias, and Heike Santen (2009). 'Trying to Catch up with the Executive Board: Regulatory Decision-Making and its Impact on CDM Performance.' In *Legal Aspects of Carbon Trading*, edited by David Freestone and Charlotte Streck. New York: Oxford University Press, 231–247.

Kuokkanen, Tuomas (2007). 'The Convention on Long-Range Transboundary Air Pollution.' In *Making Treaties Work: Human Rights, Environment and Arms Control*, edited by Geir Ulfstein, Thilo Marauhn, and Andreas Zimmermann. Cambridge, UK: Cambridge University Press, 161–178.

Lang, Winfried (1999). 'UN Principles and International Environmental Law.' 3 *Max Planck Yearbook of United Nations Law* 157–172.

Lee, Carrie, and Michael Lazarus (2011). *Bioenergy Projects and Sustainable Development: Which Project Types Offer the Greatest Benefits?* Stockholm: Stockholm Environment Institute.

Lefeber, René (2001). 'From The Hague to Bonn to Marrakesh and Beyond: A Negotiating History of the Compliance Regime under the Kyoto Protocol'. 14 *Hague Yearbook of International Law* 25–54.

Lefeber, René (2009). 'The Practice of the Compliance Committee under the Kyoto Protocol to the United Nations Framework Convention on Climate Change (2006–2007).' In *Non-Compliance Procedures and Mechanisms and the Effectiveness of International Environmental Agreements*, edited by Tullio Treves, Laura Pineschi, Attila Tanzi, Cesare Pitea, Chiara Ragni, and Francesca Romanin Jacur. The Hague: TMC Asser, 303–317.

Lefeber, René (2012). 'Climate Change and State Responsibility.' In *International Law in the Era of Climate Change*, edited by Rosemary Rayfuse and Shirley V. Scott. Cheltenham, UK: Edward Elgar, 321–349.

Lin, Jolene (2014). 'Litigating Climate Change in Asia.' 4(1) *Climate Law* 140–149.

Lohmann, Larry (2009). 'Regulatory Challenges for Financial and Carbon Markets.' 3(2) *Carbon and Climate Law Review* 161–171.

Lund, Emma (2010). 'Dysfunctional Delegation: Why the Design of the CDM's Supervisory System is Fundamentally Flawed.' 10 *Climate Policy* 277–288.

Lyster, Rosemary, Catherine MacKenzie, and Constance McDermott (eds) (2013). *Law, Tropical Forests and Carbon: The Case of REDD+*. Cambridge, UK: Cambridge University Press.

McDonald, Jan (2014). 'A Short History of Climate Adaptation Law in Australia.' 4(1–2) *Climate Law* 150–167.

MacFaul, Larry (2006). 'Developing the Climate Change Regime: The Role of Verification.' In *Verifying Treaty Compliance: Limiting Weapons of Mass Destruction and Monitoring Kyoto Protocol Provisions*, edited by Rudolf Avenhaus, Nicholas Kyriakopoulos, Michel Richard, and Gotthard Stein. New York: Springer, 171–209.

Machado-Filho, Haroldo (2012). 'Financial Mechanisms Under the Climate Regime.' In *Promoting Compliance in an Evolving Climate Regime*, edited by Jutta Brunnée, Meinhard Doelle, and Lavanya Rajamani. Cambridge, UK: Cambridge University Press, 216–239.

Machado-Filho, Haroldo (2013). 'Climate Change Policy and Legislation in Brazil.' In *Climate Change and the Law*, edited by Erkki J. Hollo, Kati Kulovesi, and Michael Mehling. Dordrecht: Springer, 639–651.

Mackenzie, Ruth (2012). 'The Role of Dispute Settlement in the Climate Regime.' In *Promoting Compliance in an Evolving Climate Regime*, edited by Jutta Brunnée, Meinhard Doelle, and Lavanya Rajamani. Cambridge, UK: Cambridge University Press, 395–417.

McPhee, John (1989). *The Control of Nature*. New York: Farrar, Straus, Giroux.

Maguire, Rowena (2013). 'Foundations of International Climate Law: Objectives, Principles and Methods.' In *Climate Change and the Law*, edited by Erkki J. Hollo, Kati Kulovesi, and Michael Mehling. Dordrecht: Springer, 83–110.

Maljean-Dubois, Sandrine, and Anne-Sophie Tabau (2012). 'From the Kyoto Compliance System to MRV: What is at Stake for the European Union?' In *Promoting Compliance in an Evolving Climate Regime*, edited by Jutta Brunnée, Meinhard Doelle, and Lavanya Rajamani. Cambridge, UK: Cambridge University Press, 317–338.

Malone, Linda A. (2011). 'Enforcing International Environmental Law Through Domestic Law Mechanisms in the United States: Civil Society Initiatives Against Global Warming.' In *Compliance and Enforcement in Environmental Law: Toward More Effective Implementation*, edited by LeRoy Paddock, Du Qun, Louis J. Kotzé, David L. Markell, Kenneth J. Markowitz, and Durwood Zaelke. Cheltenham, UK: Edward Elgar, 111–155.

Manning, Andrew C., Euan G. Nisbet, Ralph F. Keeling, and Peter S. Liss (2011). 'Greenhouse Gases in the Earth System: Setting the Agenda to 2030.' 369 *Philosophical Transactions of the Royal Society A* 1885–1890.

Marauhn, Thilo (1996). 'Towards a Procedural Law of Compliance Control in International Environmental Relations.' 56 *Zeitschrift für Ausländisches Öffentliches Recht und Völkerrecht* 696–731.

Massai, Leonardo (2009). 'Obligations of the European Community and its Member States under the Kyoto Protocol.' In *Non-Compliance Procedures and Mechanisms and the Effectiveness of International Environmental Agreements*, edited by Tullio Treves, Laura Pineschi, Attila Tanzi, Cesare Pitea, Chiara Ragni, and Francesca Romanin Jacur. The Hague: TMC Asser, 545–565.

Matthews Glenn, Jane, and José Otero (2013). 'Canada and the Kyoto Protocol: An Aesop Fable.' In *Climate Change and the Law*, edited by Erkki J. Hollo, Kati Kulovesi, and Michael Mehling. Dordrecht: Springer, 489–507.

Matthews, Robin B., Meine van Noordwijk, Eric Lambin, Patrick Meyfroidt, Joyeeta Gupta, Louis Verchot, Kristell Hergoualc'h, and Edzo Veldkamp (2014). 'Implementing REDD: Evidence on Governance, Evaluation and Impacts from the REDD-ALERT Project.' 19(6) *Mitigation and Adaptation Strategies for Global Change* 907–925.

Mehling, Michael (2012). 'Enforcing Compliance in an Evolving Climate Regime.' In *Promoting Compliance in an Evolving Climate Regime*, edited by Jutta Brunnée, Meinhard Doelle, and Lavanya Rajamani. Cambridge, UK: Cambridge University Press, 194–215.

Mehling, Michael (2013). 'Implementing Climate Governance: Instrument Choice and Interaction.' In *Climate Change and the Law*, edited by Erkki J. Hollo, Kati Kulovesi, and Michael Mehling. Dordrecht: Springer, 11–30.

Meyers, Stephen (1999). *Additionality of Emissions Reductions From Clean Development Mechanism Projects: Issues and Options for Project-Level Assessment*. Lawrence Berkeley National Laboratory, Research Paper LBNL-43704.

Michaelowa, Axel (2005). 'Determination of Baselines and Additionality for the CDM: A Crucial Element of Credibility of the Climate Regime.' In *Climate Change and Carbon Markets: A Handbook of Emission Reduction Mechanisms*, edited by Farhana Yamin. London: Earthscan, 289–303.

Michaelowa, Axel, and Pallav Purohit (2007). *Additionality Determination of Indian CDM Projects: Can Indian CDM Project Developers Outwit the CDM Executive Board?* University of Zurich, Institute for Political Science, Discussion Paper CDM-1.

Mitchell, Ronald B. (1993). 'Compliance Theory: A Synthesis.' 2(4) *Review of European Community and International Environmental Law* 327–334.

Mitchell, Ronald B. (1998). 'Sources of Transparency: Information Systems in International Regimes.' 42 *International Studies Quarterly* 109–130.

Mitchell, Ronald B. (2007). 'Compliance Theory: Compliance, Effectiveness, and Behaviour Change in International Environmental Law.' In *The Oxford Handbook of International Environmental Law*, edited by Daniel Bodansky, Jutta Brunnée, and Ellen Hey. Oxford: Oxford University Press, 893–921.

Mundaca, Luis, and Jessika Luth Richter (2014). 'Challenges for New Zealand's Carbon Market.' 3 *Nature Climate Change* 1006–1008.

Murase, Shinya, Lavanya Rajamani, et al. (2014). *Legal Principles Relating to Climate Change: Report and Draft Declaration for Consideration at the 2014 Washington Conference*. London: International Law Association. Available at <www.ila-hq.org/en/committees/index.cfm/cid/1029>.

Murdiyarso, Daniel, Maria Brockhaus, William D. Sunderlin, and Lou Verchot (2012). 'Some Lessons Learned From the First Generation of REDD+ Activities.' 4 *Current Opinion in Environmental Sustainability* 678–685.

Nachmany, Michal, Sam Fankhauser, Terry Townshend, Murray Collins, Tucker Landesman, Adam Matthews, Carolina Pavese, Katharina Rietig, Philip Schleifer, and Joana Setzer (2014). *The GLOBE Climate Legislation Study: A Review of Climate Change Legislation in 66 Countries (4th edn)*. London: GLOBE International and Grantham Research Institute.

Nagle, John Copeland (2010). 'Climate Exceptionalism.' 40 *Environmental Law* 53–88.

Nakhooda, Smita, Taryn Fransen, Takeshi Kuramochi, Alice Caravani, Annalisa Prizzon, Noriko Shimizu, Helen Tilley, Aidy Halimanjaya, and Bryn Welham (2013). *Mobilising International Climate Finance: Lessons from the Fast-Start Finance Period*. Washington DC: World Resources Institute.

Narain, Urvashi, Sergio Margulis, and Timothy Essam (2011). 'Estimating Costs of Adaptation to Climate Change.' 11(3) *Climate Policy* 1001–1019.

Nature Editorial (2012). 'Gas and Air.' 482 *Nature* 131.

Nature Editorial (2013). 'The Final Assessment.' 501 *Nature* 281.

Neeff, Till, and Francisco Ascui (2009). 'Lessons From Carbon Markets for Designing an Effective REDD Architecture.' 9 *Climate Policy* 306–315.

Netto, Maria, and Kai-Uwe Barani Schmidt (2005). 'CDM Project Cycle and the Role of the UNFCCC Secretariat.' In *Legal Aspects of Implementing the Kyoto Protocol Mechanisms: Making Kyoto Work*, edited by David Freestone and Charlotte Streck. New York: Oxford University Press, 175–190.

Newell, Peter, and Matthew Paterson (2010). *Climate Capitalism: Global Warming and the Transformation of the Global Economy*. Cambridge, UK: Cambridge University Press.

Nollkaemper, André (2002). 'Compliance Control in International Environmental Law: Traversing the Limits of the National Legal Order'. 13 *Yearbook of International Environmental Law* 165–186.

Nussbaumer, Patrick (2009). 'On the Contribution of Labelled Certified Emission Reductions to Sustainable Development: A Multi-Criteria Evaluation of CDM Projects.' 37(1) *Energy Policy* 91–101.

Oberthür, Sebastian (2014). 'Options for a Compliance Mechanism in a 2015 Climate Agreement.' 4(1–2) *Climate Law* 30–49.

Oberthür, Sebastian, and René Lefeber (2010). 'Holding Countries to Account: The Kyoto Protocol's Compliance System Revisited after Four Years of Experience.' 1(1) *Climate Law* 133–158.

O'Connell, Mary Ellen (1995). 'Enforcement and the Success of International Environmental Law.' 3(1) *Indiana Journal of Global Legal Studies* 47–64.

O'Connell, Mary Ellen (2000). 'The Role of Soft Law in a Global Order.' In *Commitment and Compliance: The Role of Non-Binding Norms in the International Legal System*, edited by Dinah Shelton. Oxford: Oxford University Press, 100–118.

Olbrisch, Susanne, Erik Haites, Matthew Savage, Pradeep Dadhich, and Manish Kumar Shrivastava (2011). 'Estimates of Incremental Investment for and Cost of Mitigation Measures in Developing Countries.' 11(3) *Climate Policy* 970–986.

Olsen, Karen Holm (2007). 'The Clean Development Mechanism's Contribution to Sustainable Development: A Review of the Literature.' 84 *Climatic Change* 59–73.

Olsen, Karen Holm, and Jørgen Fenhann (2008). 'Sustainable Development Benefits of Clean Development Mechanism Projects.' 36(8) *Energy Policy* 2819–2830.

Osofsky, Hari (2012). 'Climate Change and Dispute Resolution Processes.' In *International Law in the Era of Climate Change*, edited by Rosemary Rayfuse and Shirley V. Scott. Cheltenham, UK: Edward Elgar, 350–370.

Peeters, Marjan (2006). 'Enforcement of the EU Greenhouse Gas Emissions Trading Scheme.' In *EU Climate Change Policy: The Challenge of New Regulatory Initiatives*, edited by Marjan Peeters and Kurt Deketelaere. Cheltenham, UK: Edward Elgar, 169–187.

Peeters, Marjan, and Thomas Schomerus (2014). 'Modifying Our Society With Law: The Case of EU Renewable-Energy Law.' 4(1–2) *Climate Law* 131–139.

Peters, Glen P., Gregg Marland, Corinne Le Quéré, Thomas Boden, Josep G. Canadell, and Michael R. Raupach (2012). 'Rapid Growth in CO_2 Emissions after the 2008–2009 Global Financial Crisis.' 2 *Nature Climate Change* 2–4.

Peters, Glen P., Jan C. Minx, Christopher L. Weber, and Ottmar Edenhofer (2011). 'Growth in Emission Transfers via International Trade From 1990 to 2008.' 108(21) *PNAS* 8903–8908.

Polycarp, Clifford, Louise Brown, and Xing Fu-Bertaux (2013). *Mobilizing Climate Investment: The Role of International Climate Finance in Creating Readiness for Scaled-up Low-Carbon Energy*. Washington DC: World Resources Institute.

Posner, Eric A. and David Weisbach (2010). *Climate Change Justice*. Princeton, NJ: Princeton University Press.

Porter, Jean (2007). 'Custom, Ordinance and Natural Right in Gratian's *Decretum*.' In *The Nature of Customary Law*, edited by Amanda Perreau-Saussine and James Bernard Murphy. New York: Cambridge University Press, pp. 79–100.

Potter, Christopher, Peggy Gross, Steven Klooster, Matthew Fladeland, and Vanessa Genovese (2008). 'Storage of Carbon in U.S. Forests Predicted from Satellite Data, Ecosystem Modeling, and Inventory Summaries.' 90 *Climatic Change* 269–282.

Rajamani, Lavanya (2012). 'Developing Countries and Compliance in the Climate Regime.' In *Promoting Compliance in an Evolving Climate Regime*, edited by Jutta Brunnée, Meinhard Doelle, and Lavanya Rajamani. Cambridge, UK: Cambridge University Press, 367–394.

Rastogi, Patodia (2013). 'India's Evolving Climate Change Strategy.' In *Climate Change and the Law*, edited by Erkki J. Hollo, Kati Kulovesi, and Michael Mehling. Dordrecht: Springer, 605–618.

Raustiala, Kal, and Anne-Marie Slaughter (2002). 'International Law, International Relations and Compliance.' In *Handbook of International Relations*, edited by Walter Carlsnaes, Thomas Risse, and Beth A. Simmons. London: SAGE, 538–558.

Redgwell, Catherine (2007). 'National Implementation.' In *The Oxford Handbook of International Environmental Law*, edited by Daniel Bodansky, Jutta Brunnée, and Ellen Hey. Oxford: Oxford University Press, 922–946.

Redgwell, Catherine (2012a). 'Climate Change and International Environmental Law.' In *International Law in the Era of Climate Change*, edited by Rosemary Rayfuse and Shirley V. Scott. Cheltenham, UK: Edward Elgar, 118–146.

Redgwell, Catherine (2012b). 'Facilitation of Compliance.' In *Promoting Compliance in an Evolving Climate Regime*, edited by Jutta Brunnée, Meinhard Doelle, and Lavanya Rajamani. Cambridge, UK: Cambridge University Press, 177–193.

Reeve, Rosalind (2002). *Policing International Trade in Endangered Species: The CITES Treaty and Compliance*. London: The Royal Institute of International Affairs/Earthscan.

Reeve, Rosalind (2007). 'The Convention on International Trade in Endangered Species of Wild Fauna and Flora (CITES).' In *Making Treaties Work: Human Rights, Environment and Arms Control*, edited by Geir Ulfstein, Thilo Marauhn, and Andreas Zimmermann. Cambridge, UK: Cambridge University Press, 134–160.

Reid, Colin T. (2013). 'Climate Law in the United Kingdom.' In *Climate Change and the Law*, edited by Erkki J. Hollo, Kati Kulovesi, and Michael Mehling. Dordrecht: Springer, 537–549.

Ringius, Lasse, Asbjørn Torvanger, and Arild Underdal (2002). 'Burden Sharing and Fairness Principles in International Climate Policy.' 2 *International Environmental Agreements* 1–22.

Rivera-Batiz, Francisco L. (2002). 'Democracy, Governance, and Economic Growth: Theory and Evidence.' 6(2) *Review of Development Economics* 225–247.

Rivier, Leonard et al. (2010). 'European CO_2 Fluxes from Atmospheric Inversions Using Regional and Global Transport Models.' 103 *Climatic Change* 93–115.

Romijn, Erika, Martin Herold, Lammert Kooistra, Daniel Murdiyarso, and Louis Verchot (2012). 'Assessing Capacities of Non-Annex I Countries for National Forest Monitoring in the Context of REDD+.' 19–20 *Environmental Science and Policy* 33–48.

Rose, Gregory (2011). 'Interlinkages Between Multi-Lateral Environmental Agreements: International Compliance Cooperation.' In *Compliance and Enforcement in Environmental Law: Toward More Effective Implementation*, edited by LeRoy Paddock, Du Qun, Louis J. Kotzé, David L. Markell, Kenneth J. Markowitz, and Durwood Zaelke. Cheltenham, UK: Edward Elgar, 3–33.

Rose, Gregory, and Lal Kurukulasuriya (2007). *Compliance Mechanisms under Selected Multilateral Environmental Agreements*. Nairobi: United Nations Environment Programme.

Ruhl, J.B., and James Salzman (2013). 'Climate Change Meets the Law of the Horse.' 62(4) *Duke Law Journal* 975–1027.

Rypdal, Kristin, Frode Stordal, Jan Fuglestvedt, and Terje Berntsen (2005). 'Introducing Top-Down Methods in Assessing Compliance with the Kyoto Protocol.' 5 *Climate Policy* 393–405.

Sand, Peter H. (1996). 'Institution-Building to Assist Compliance with International Environmental Law: Perspectives.' 56 *Zeitschrift für Ausländisches Öffentliches Recht und Völkerrecht* 774–795.

Sands, Philippe (2006). 'Non-Compliance and Dispute Settlement.' In *Ensuring Compliance with Multilateral Environmental Agreements*, edited by Ulrich Beyerlin, Peter-Tobias Stoll, and Rüdiger Wolfrum. Leiden: Martinus Nijhoff, 353–358.

Sbragia, Alberta M., and Philipp M. Hildebrand (1998). 'The European Union and Compliance: A Story in the Making.' In *Engaging Countries: Strengthening Compliance with International Environmental Accords*, edited by Edith Brown Weiss and Harold K. Jacobson. Cambridge, MA: MIT Press, 215–252.

Schmidt, Gavin A. (2014). 'On Scientists and Advocacy.' 344 *Science* 256.

Schneider, Lambert (2007). *Is the CDM Fulfilling Its Environmental and Sustainable Development Objectives? An Evaluation of the CDM and Options for Improvement*. Freiburg, Germany: Öko-Institut.

Schneider, Lambert (2009). 'Assessing the Additionality of CDM Projects: Practical Experience and Lessons Learned.' 9 *Climate Policy* 242–254.

Schneider, Lambert (2011). 'Perverse Incentives Under the CDM: An Evaluation of HFC-23 Destruction Projects.' 11 *Climate Policy* 851–864.

Scholz, Sebastian, and Ian Noble (2005). 'Generation of Sequestration Credits under the CDM.' In *Legal Aspects of Implementing the Kyoto Protocol Mechanisms: Making Kyoto Work*, edited by David Freestone and Charlotte Streck. New York: Oxford University Press, 265–280.

Schueler, Ben (2011). 'Governmental Liability: An Incentive for Appropriate Adaptation?' In *Climate Change Liability*, edited by Michael Faure and Marjan Peeters. Cheltenham, UK: Edward Elgar, 237–251.

Silva-Castañeda, Laura (2012). 'A Forest of Evidence: Third-Party Certification and Multiple Forms of Proof—A Case Study of Oil Palm Plantations in Indonesia.' 29(3) *Agriculture and Human Values* 361–370.

Sindico, Francesco (2012). 'Post-2012 Compliance and Carbon Markets.' In *Promoting Compliance in an Evolving Climate Regime*, edited by Jutta Brunnée, Meinhard Doelle, and Lavanya Rajamani. Cambridge, UK: Cambridge University Press, 240–261.

Smirnov, D.Y., A.G. Kabanets, B.J. Milakovsky, E.A. Lepeshkin, and D.V. Sychikov (2013). *Illegal Logging in the Russian Far East: Global Demand and Taiga Destruction*. Moscow: WWF.

Smith, Joel B., Thea Dickinson, Joseph D.B. Donahue, Ian Burton, Erik Haites, Richard J.T. Klein, and Anand Patwardhan (2011). 'Development and Climate Change Adaptation Funding: Coordination and Integration.' 11(3) *Climate Policy* 987–1000.

Soltau, Friedrich (2009). *Fairness in International Climate Change Law and Policy*. Cambridge: Cambridge University Press.

Soudani, Kamel, and Christophe François (2014). 'A Green Illusion.' 506 *Nature* 165.

Spier, Jaap (2011). 'High Noon: Prevention of Climate Damage as the Primary Goal of Liability?' In *Climate Change Liability*, edited by Michael Faure and Marjan Peeters. Cheltenham, UK: Edward Elgar, 47–51.

Stiglitz, Joseph E. (2002). 'Participation and Development: Perspectives from the Comprehensive Development Paradigm.' 6(2) *Review of Development Economics* 163–182.

Strauss, Andrew (2009). 'Climate Change Litigation: Opening the Door to the International Court of Justice.' In *Adjudicating Climate Change: State, National, and*

International Approaches, edited by William C.G. Burns and Hari M. Osofsky. New York: Cambridge University Press, 334–356.

Streck, Charlotte (2001). 'The Global Environment Facility—A Role Model for International Governance?' 1 *Global Environmental Politics* 71–94.

Streck, Charlotte, and Jolene Lin (2009). 'Mobilising Finance for Climate Change Mitigation: Private Sector Involvement in International Carbon Finance Mechanisms.' 10(1) *Melbourne Journal of International Law* 70–101.

Subbarao, Srikanth, and Bob Lloyd (2011). 'Can the Clean Development Mechanism (CDM) Deliver?' 39(3) *Energy Policy* 1600–1611.

Sugiyama, Taishi, and Axel Michaelowa (2001). 'Reconciling the Design of CDM With Inborn Paradox of Additionality Concept.' 1(1) *Climate Policy* 75–83.

Sutter, Christoph, and Juan Carlos Parreño (2007). 'Does the Current Clean Development Mechanism (CDM) Deliver its Sustainable Development Claim? An Analysis of Officially Registered CDM Projects.' 84 *Climatic Change* 75–90.

Swart, Rob, Peter Bergamaschi, Tinus Pulles, and Frank Raes (2007). 'Are National Greenhouse Gas Emissions Reports Scientifically Valid?' 7 *Climate Policy* 535–538.

Széll, Patrick (1995). 'The Development of Multilateral Mechanisms for Monitoring Compliance.' In *Sustainable Development and International Law*, edited by Winfried Lang. London: Graham and Trotman, 97–109.

Trachtman, Joel P. (1999). 'Bananas, Direct Effect and Compliance.' 10(4) *European Journal of International Law* 655–678.

Tung, Christopher (2013). 'Sustainable Development and Climate Policy and Law in China.' In *Climate Change and the Law*, edited by Erkki J. Hollo, Kati Kulovesi, and Michael Mehling. Dordrecht: Springer, 597–603.

Ulfstein, Geir (2007). 'Dispute Resolution, Compliance Control and Enforcement in International Environmental Law.' In *Making Treaties Work: Human Rights, Environment and Arms Control*, edited by Geir Ulfstein, Thilo Marauhn, and Andreas Zimmermann. Cambridge, UK: Cambridge University Press, 115–133.

Ulfstein, Geir (2012). 'Depoliticizing Compliance.' In *Promoting Compliance in an Evolving Climate Regime*, edited by Jutta Brunnée, Meinhard Doelle, and Lavanya Rajamani. Cambridge, UK: Cambridge University Press, 418–433.

Ulfstein, Geir, and Jacob Werksman (2005). 'The Kyoto Compliance System: Towards Hard Enforcement.' In *Implementing the Climate Regime: International Compliance*, edited by Jon Hovi, Olav Stokke, and Geir Ulfstein. London: Earthscan, 39–62.

Urbinati, Sabrina (2009). 'Procedures and Mechanisms Relating to Compliance under the 1997 Kyoto Protocol to the 1992 United Nations Framework Convention on Climate Change.' In *Non-Compliance Procedures and Mechanisms and the Effectiveness of International Environmental Agreements*, edited by Tullio Treves, Laura Pineschi, Attila Tanzi, Cesare Pitea, Chiara Ragni, and Francesca Romanin Jacur. The Hague: TMC Asser, 63–84.

Van Asselt, Harro (2013). 'Managing the Fragmentation of International Climate Law.' In *Climate Change and the Law*, edited by Erkki J. Hollo, Kati Kulovesi, and Michael Mehling. Dordrecht: Springer, 329–357.

Vandenbergh, Michael P., and Anne C. Steinemann (2007). 'The Carbon-Neutral Individual.' 82 *New York University Law Review* 1673–1745.

Van Dijk, Chris (2011). 'Civil Liability for Global Warming in the Netherlands.' In *Climate Change Liability*, edited by Michael Faure and Marjan Peeters. Cheltenham, UK: Edward Elgar, 206–226.

Van Vliet, Oscar, André Faaij, and Carel Dieperink (2003). 'Forestry Projects under the Clean Development Mechanism?' 61 *Climatic Change* 123–156.

Victor, David G. (1998). 'The Operation and Effectiveness of the Montreal Protocol's Non-Compliance Procedure.' In *The Implementation and Effectiveness of International Environmental Commitments: Theory and Practice*, edited by David G. Victor, Kal Raustiala, and Eugene B. Skolnikoff. Cambridge, MA: MIT Press, 137–176.

Victor, David G. (1999). 'Enforcing International Law: Implications for an Effective Global Warming Regime.' 10 *Duke Environmental Law and Policy Forum* 147–184.

Victor, David G. (2008). 'On the Regulation of Geoengineering.' 24(2) *Oxford Review of Economic Policy* 322–336.

Vihma, Antto (2013). 'Analyzing Soft Law and Hard Law in Climate Change.' In *Climate Change and the Law*, edited by Erkki J. Hollo, Kati Kulovesi, and Michael Mehling. Dordrecht: Springer, 143–164.

Virgilio, Nicole R., Sarene Marshall, Olaf Zerbock, and Christopher Holmes (2010). *Reducing Emissions from Deforestation and Degradation (REDD): A Casebook of on-the-Ground Experience*. The Nature Conservancy.

Voigt, Christina (2009). 'The Deadlock of the Clean Development Mechanism: Caught between Sustainability, Environmental Integrity and Economic Efficiency.' In *Climate Law and Developing Countries: Legal and Policy Changes for the World Economy*, edited by Benjamin J. Richardson, Yves Le Bouthillier, Heather McLeod-Kilmurray, and Stepan Wood. Cheltenham, UK: Edward Elgar, 235–261.

Wang, Qiang (2014). 'China Should Aim For a Total Cap on Emissions.' 512 *Nature* 115.

Wang, Xueman and Glenn Wiser (2002). 'The Implementation and Compliance Regimes under the Climate Change Convention and its Kyoto Protocol.' 11(2) *Review of European Community and International Environmental Law* 181–198.

Wara, Michael W., and David G. Victor (2008). *A Realistic Policy on International Carbon Offsets*. Program on Energy and Sustainable Development, Stanford University, Working Paper No. 74.

Weiss, Edith Brown (1999). 'Understanding Compliance with International Environmental Agreements: The Baker's Dozen Myths.' 32(5) *University of Richmond Law Review* 1555–1589.

Werksman, Jacob (1996). 'Compliance and Transition: Russia's Non-Compliance Tests the Ozone Regime.' 56 *Zeitschrift für Ausländisches Öffentliches Recht und Völkerrecht* 750–773.

Werksman, Jacob (1999). 'Compliance and the Kyoto Protocol: Building a Backbone into a "Flexible" Regime.' 9 *Yearbook of International Environmental Law* 48–101.

Werksman, Jacob (2005). 'The Negotiation of a Kyoto Compliance System.' In *Implementing the Climate Regime: International Compliance*, edited by Jon Hovi, Olav Stokke, and Geir Ulfstein. London: Earthscan, 17–37.

Werksman, Jacob (2012). 'Compliance and the Use of Trade Measures.' In *Promoting Compliance in an Evolving Climate Regime*, edited by Jutta Brunnée, Meinhard Doelle, and Lavanya Rajamani. Cambridge, UK: Cambridge University Press, 262–285.

Wettestad, Jørgen (2007). 'Monitoring and Verification.' In *The Oxford Handbook of International Environmental Law*, edited by Daniel Bodansky, Jutta Brunnée, and Ellen Hey. Oxford: Oxford University Press, 974–994.

Wiersema, Annecoos (2009). 'The New International Law-Makers? Conferences of the Parties to Multilateral Environmental Agreements.' 31(1) *Michigan Journal of International Law* 231–287.

Wiersema, Annecoos (2013). *Climate Change, Forests, and International Law: REDD's Descent into Irrelevance*. <http://works.bepress.com/annecoos_wiersema/2>.

Winiwarter, Wilfried (2007). 'National Greenhouse Gas Inventories: Understanding Uncertainties versus Potential for Improving Reliability.' 7(4–5) *Water, Air, and Soil Pollution: Focus* 443–450.

Wolfrum, Rüdiger (1998). 'Means of Ensuring Compliance With and Enforcement of International Environmental Law.' 272 *Recueil des cours* 9–154.

Wolfrum, Rüdiger (2000). 'International Environmental Law: Purposes, Principles and Means of Ensuring Compliance.' In *International, Regional and National Environmental Law*, edited by Fred L. Morrison and Rüdiger Wolfrum. The Hague: Kluwer Law International, 3–70.

Yamin, Farhana, and Joanna Depledge (2004). *The International Climate Change Regime: A Guide to Rules, Institutions and Procedures*. Cambridge: Cambridge University Press.

Yamineva, Yulia (2013). 'Climate Law and Policy in Russia: A Peasant Needs Thunder to Cross Himself and Wonder.' In *Climate Change and the Law*, edited by Erkki J. Hollo, Kati Kulovesi, and Michael Mehling. Dordrecht: Springer, 551–566.

Yamineva, Yulia, and Kati Kulovesi (2013). 'The New Framework for Climate Finance Under the United Nations Framework Convention on Climate Change: A Breakthrough or an Empty Promise?' In *Climate Change and the Law*, edited by Erkki J. Hollo, Kati Kulovesi, and Michael Mehling. Dordrecht: Springer, 191–223.

Yoshida, O. (1999). 'Soft Enforcement of Treaties: The Montreal Protocol's Noncompliance Procedure and the Functions of Internal International Institutions.' 10(1) *Colorado Journal of International Environmental Law and Policy* 95–141.

Young, Oran R. (1994). *International Governance: Protecting the Environment in a Stateless Society*. Ithaca, NY: Cornell University Press.

Zahar, Alexander (2010a). 'Does Self-Interest Skew State Reporting of Greenhouse Gas Emissions? A Preliminary Analysis Based on the First Verified Emissions Estimates under the Kyoto Protocol.' 1(2) *Climate Law* 313–324.

Zahar, Alexander (2010b). 'Verifying Greenhouse Gas Emissions of Annex I Parties: Methods We Have and Methods We Want.' 1(3) *Climate Law* 409–427.

Zahar, Alexander (2012), 'The Climate Change Regime.' In *The Routledge Handbook of International Environmental Law*, edited by Shawkat Alam, Jahid Hossain Bhuiyan, Tareq M.R. Chowdhury, and Erika J. Techera. Oxford, UK: Routledge, 349–373.

Zahar, Alexander (2014). 'International Environmental Institutions.' In *Oxford Bibliographies in International Law*, edited by Tony Carty. New York: Oxford University Press (online chapter).

Zahar, Alexander, Jacqueline Peel, and Lee Godden (2012). *Australian Climate Law in Global Context*. Cambridge, UK: Cambridge University Press.

Zhang, Junjie, and Can Wang (2011). 'Co-Benefits and Additionality of the Clean Development Mechanism: An Empirical Analysis.' 62 *Journal of Environmental Economics and Management* 140–154.

2. Intergovernmental and NGO reports

Environmental Defense Fund (2013). *Annual Report 2013*.

Environmental Investigation Agency (2011). *Crossroads: The Illicit Timber Trade Between Laos and Vietnam*. Available from <http://www.eia-international.org/category/reports>.

Food and Agriculture Organization (2010). *Global Forest Resources Assessment 2010: Main Report*.

Food and Agriculture Organization and International Tropical Timber Organization (2011). *The State of Forests in the Amazon Basin, Congo Basin and Southeast Asia: A Report Prepared for the Summit of the Three Rainforest Basins, Brazzaville, Republic of Congo, 31 May – 3 June 2011.*

Food and Agriculture Organization, UN Development Programme, and UN Environment Programme (2011). *UN-REDD Programme, May 2011 Newsletter* (vol. 18).

Food and Agriculture Organization, UN Development Programme, and UN Environment Programme (2011), *UN-REDD Programme, June 2011 Newsletter* (vol. 19).

Forest Europe, UN Economic Commission for Europe, and Food and Agriculture Organization (2011). *State of Europe's Forests 2011: Status and Trends in Sustainable Forest Management in Europe.* Ministerial Conference on the Protection of Forests in Europe.

Friends of the Earth International (2012). *Annual Report 2012.*

Global Environment Facility (2013). *Final Report of the Fifth Overall Performance Study of the GEF: At Crossroads for Higher Impact.* GEF/R.6/17.

Global Environment Facility (2013). *Report of the Global Environment Facility to the Nineteenth Session of the Conference of the Parties to the United Nations Framework Convention on Climate Change.* FCCC/CP/2013/3.

Global Forest Observations Initiative (2014). *Integrating Remote-Sensing and Ground-Based Observations for Estimation of Emissions and Removals of Greenhouse Gases in Forests: Methods and Guidance from the Global Forest Observations Initiative.* Geneva: Group on Earth Observations.

Governing Council of the United Nations Environment Programme (2011). *Environment in the United Nations System.* UNEP/GC.26/INF/23.

Greenpeace International (2012). *Annual Report 2012.*

Intergovernmental Panel on Climate Change (1996). *Revised 1996 IPCC Guidelines for National Greenhouse Gas Inventories.*

Intergovernmental Panel on Climate Change (2000). *Good Practice Guidance and Uncertainty Management in National Greenhouse Gas Inventories.*

Intergovernmental Panel on Climate Change (2003). *Good Practice Guidance for Land Use, Land-Use Change and Forestry.*

Intergovernmental Panel on Climate Change (2006). *IPCC Guidelines for National Greenhouse Gas Inventories.*

Intergovernmental Panel on Climate Change (2007). *Climate Change 2007: Mitigation: Contribution of Working Group III to the Fourth Assessment Report of the Intergovernmental Panel on Climate Change.* New York: Cambridge University Press.

Intergovernmental Panel on Climate Change (2013). *Climate Change 2013: The Physical Science Basis: Working Group I Contribution to the Fifth Assessment Report of the Intergovernmental Panel on Climate Change.* New York: Cambridge University Press.

Intergovernmental Panel on Climate Change (2013). *Climate Change 2013: The Physical Science Basis: Working Group I Contribution to the Fifth Assessment Report of the Intergovernmental Panel on Climate Change: Summary for Policymakers.*

Intergovernmental Panel on Climate Change (2014a). *Climate Change 2014: Impacts, Adaptation, and Vulnerability: Working Group II Contribution to the Fifth Assessment Report of the Intergovernmental Panel on Climate Change: Summary for Policymakers.*

Intergovernmental Panel on Climate Change (2014b). *Climate Change 2014: Mitigation of Climate Change: Working Group III Contribution to the Fifth Assessment Report of the Intergovernmental Panel on Climate Change (Final Draft).*

Intergovernmental Panel on Climate Change (2014c). *Climate Change 2014: Mitigation of Climate Change: Working Group III Contribution to the Fifth Assessment Report of the Intergovernmental Panel on Climate Change: Summary for Policymakers.*
International Energy Agency (2009). *World Energy Outlook 2009.* Paris: IEA.
International Energy Agency (2010). CO_2 *Emissions from Fuel Combustion: Highlights.*
International Energy Agency (2013). *World Energy Outlook 2013.* Paris: IEA.
International Institute for Sustainable Development (2010). 'SB 32 and AWG Highlights: Wednesday, 9 June 2010.' 12(470) *Earth Negotiations Bulletin* 1–4.
International Institute for Sustainable Development (2011a). 'SB 34 and AWG Highlights: Tuesday, 14 June 2011.' 12(510) *Earth Negotiations Bulletin* 1–4.
International Institute for Sustainable Development (2011b). 'AWG-LCA 14 and AWG-KP 16 Highlights: Wednesday, 5 October 2011.' 12(519) *Earth Negotiations Bulletin* 1–4.
International Institute for Sustainable Development (2013a). 'Summary of the Bonn Climate Change Conference: 29 April – 3 May 2013.' 12(568) *Earth Negotiations Bulletin* 1–19.
International Institute for Sustainable Development (2013b). 'Warsaw Highlights: Tuesday, 12 November 2013.' 12(585) *Earth Negotiations Bulletin* 1–4.
International Institute for Sustainable Development (2013c). 'Warsaw Highlights: Friday, 15 November 2013.' 12(588) *Earth Negotiations Bulletin* 1–4.
International Institute for Sustainable Development (2013d). 'Warsaw Highlights: Thursday, 21 November 2013', 12(593) *Earth Negotiations Bulletin* 1–4.
International Institute for Sustainable Development (2014). 'Summary of the Bonn Climate Change Conference: 4–15 June 2014.' 12(598) *Earth Negotiations Bulletin* 1–36.
Mo Ibrahim Foundation (2007–). 'The Ibrahim Index.' <www.moibrahimfoundation.org/iiag/>.
NASA Jet Propulsion Laboratory. 'OCO-2: Orbiting Carbon Observatory.' <http://oco.jpl.nasa.gov/>.
National Research Council, United States (2010). *Verifying Greenhouse Gas Emissions: Methods to Support International Climate Agreements.*
Rainforest Alliance (2012). *Annual Report 2012.*
Sierra Club Foundation (2012). *Annual Report 2012.*
UN Environment Programme (2013). *The Emissions Gap Report 2013: A UNEP Synthesis Report.*
United Nations Department of Economic and Social Affairs (2013). *World Population Prospects: The 2012 Revision. Volume I: Comprehensive Tables.* ST/ESA/SER.A/336.
UN-REDD (2011). *Report of the Sixth Policy Board Meeting.*
UN-REDD (2013). *Sharing National Experiences in Strengthening Transparency, Accountability and Integrity for REDD+.*
World Meteorological Organization (2013). *Greenhouse Gas Bulletin No. 9.* WMO.
World Meteorological Organization (2014). 'World Data Centre for Greenhouse Gases.' <http://ds.data.jma.go.jp/gmd/wdcgg/>.
World Resources Institute (2012). *2011–2012 Annual Report.*

3. Decisions of state parties: FCCC

Decision 2/CP.1 (1995). *Review of First Communications from the Parties Included in Annex I to the Convention.* FCCC/CP/1995/7/Add.1.

Decision 4/CP.1 (1995). *Methodological Issues.* FCCC/CP/1995/7/Add.1.

Decision 11/CP.4 (1998). *National Communications From Parties Included in Annex I to the Convention.* FCCC/CP/1998/16/Add.1.

Decision 3/CP.5 (1999). *Guidelines for the Preparation of National Communications by Parties Included in Annex I to the Convention, Part I: UNFCCC Reporting Guidelines on Annual Inventories.* FCCC/CP/1999/6/Add.1.

Decision 4/CP.5 (1999). *Guidelines for the Preparation of National Communications by Parties Included in Annex I to the Convention, Part II: UNFCCC Reporting Guidelines on National Communications.* FCCC/CP/1999/6/Add.1.

Decision 6/CP.5 (1999). *Guidelines for the Technical Review of Greenhouse Gas Inventories from Parties Included in Annex I to the Convention.* FCCC/CP/1999/6/Add.1.

Decision 5/CP.7 (2001). *Implementation of Article 4, Paragraphs 8 and 9, of the Convention.* FCCC/CP/2001/13/Add.1.

Decision 10/CP.7 (2001). *Funding Under the Kyoto Protocol.* FCCC/CP/2001/13/Add.1.

Decision 27/CP.7 (2001). *Guidance to an Entity Entrusted With the Operation of the Financial Mechanism of the Convention, for the Operation of the Least Developed Countries Fund.* FCCC/CP/2001/13/Add.4.

Decision 28/CP.7 (2001). *Guidelines for the Preparation of National Adaptation Programmes of Action.* FCCC/CP/2001/13/Add.4.

Decision 17/CP.8 (2002). *Guidelines for the Preparation of National Communications from Parties Not Included in Annex I to the Convention.* FCCC/CP/2002/7/Add.2.

Decision 18/CP.8 (2002). *Guidelines for the Preparation of National Communications by Parties Included in Annex I to the Convention, Part I: UNFCCC Reporting Guidelines on Annual Inventories.* FCCC/CP/2002/7/Add.2.

Decision 19/CP.8 (2002). *Guidelines for the Technical Review of Greenhouse Gas Inventories from Parties Included in Annex I to the Convention.* FCCC/CP/2002/7/Add.2.

Decision 1/CP.13 (2007). *Bali Action Plan.* FCCC/CP/2007/6/Add.1.

Decision 2/CP.13 (2007). *Reducing Emissions from Deforestation in Developing Countries: Approaches to Stimulate Action.* FCCC/CP/2007/6/Add.1.

Decision 10/CP.13 (2007). *Compilation and Synthesis of Fourth National Communications.* FCCC/CP/2007/6/Add.1.

Decision 4/CP.15 (2009). *Methodological Guidance for Activities Relating to Reducing Emissions from Deforestation and Forest Degradation and the Role of Conservation, Sustainable Management of Forests and Enhancement of Forest Carbon Stocks in Developing Countries.* FCCC/CP/2009/11/Add.1.

Decision 1/CP.16 (2010). *The Cancun Agreements: Outcome of the Work of the Ad Hoc Working Group on Long-Term Cooperative Action under the Convention.* FCCC/CP/2010/7/Add.1.

Decision 2/CP.16 (2010). *Fourth Review of the Financial Mechanism.* FCCC/CP/2010/7/Add.2.

Decision 2/CP.17 (2011). *Outcome of the Work of the Ad Hoc Working Group on Long-Term Cooperative Action under the Convention.* FCCC/CP/2011/9/Add.1.

Decision 3/CP.17 (2011). *Launching the Green Climate Fund.* FCCC/CP/2011/9/Add.1.

Decision 12/CP.17 (2011). *Guidance on Systems for Providing Information on How Safeguards Are Addressed and Respected and Modalities Relating to Forest Reference Emission Levels and Forest Reference Levels as Referred to in Decision 1/CP.16, Appendix I*. FCCC/CP/2011/9/Add.2.

Decision 1/CP.18 (2012). *Agreed Outcome Pursuant to the Bali Action Plan*. FCCC/CP/2012/8/Add.1.

Decision 1/CP.19 (2013). *Further Advancing the Durban Platform*. FCCC/CP/2013/10/Add.1.

Decision 2/CP.19 (2013). *Warsaw International Mechanism for Loss and Damage Associated with Climate Change Impacts*. FCCC/CP/2013/10/Add.1.

Decision 3/CP.19 (2013). *Long-Term Climate Finance*. FCCC/CP/2013/10/Add.1.

Decision 4/CP.19 (2013). *Report of the Green Climate Fund to the Conference of the Parties and Guidance to the Green Climate Fund*. FCCC/CP/2013/10/Add.1.

Decision 5/CP.19 (2013). *Arrangements between the Conference of the Parties and the Green Climate Fund*. FCCC/CP/2013/10/Add.1.

Decision 6/CP.19 (2013). *Report of the Global Environment Facility to the Conference of the Parties and Guidance to the Global Environment Facility*. FCCC/CP/2013/10/Add.1.

Decision 9/CP.19 (2013). *Work Programme on Results-Based Finance to Progress the Full Implementation of the Activities Referred to in Decision 1/CP.16, Paragraph 70*. FCCC/CP/2013/10/Add.1.

Decision 11/CP.19 (2013). *Modalities for National Forest Monitoring Systems*. FCCC/CP/2013/10/Add.1.

Decision 13/CP.19 (2013). *Guidelines and Procedures for the Technical Assessment of Submissions from Parties on Proposed Forest Reference Emission Levels and/or Forest Reference Levels*. FCCC/CP/2013/10/Add.1.

Decision 14/CP.19 (2013). *Modalities for Measuring, Reporting and Verifying*. FCCC/CP/2013/10/Add.1.

Decision 16/CP.19 (2013). *Work of the Adaptation Committee*. FCCC/CP/2013/10/Add.2.

Decision 19/CP.19 (2013). *Work of the Consultative Group of Experts on National Communications from Parties Not Included in Annex I to the Convention*. FCCC/CP/2013/10/Add.2.

Decision 20/CP.19 (2013). *Composition, Modalities and Procedures of the Team of Technical Experts under International Consultation and Analysis*. FCCC/CP/2013/10/Add.2.

Decision 21/CP.19 (2013). *General Guidelines for Domestic Measurement, Reporting and Verification of Domestically Supported Nationally Appropriate Mitigation Actions by Developing Country Parties*. FCCC/CP/2013/10/Add.2.

Decision 23/CP.19 (2013). *Work Programme on the Revision of the Guidelines for the Review of Biennial Reports and National Communications, Including National Inventory Reviews, for Developed Country Parties*. FCCC/CP/2013/10/Add.2.

Decision 24/CP.19 (2013). *Revision of the UNFCCC Reporting Guidelines on Annual Inventories for Parties Included in Annex I to the Convention*. FCCC/CP/2013/10/Add.3.

4. Decisions of state parties: Kyoto Protocol

Decision 2/CMP.1 (2005). *Principles, nature and scope of the mechanisms pursuant to Articles 6, 12 and 17 of the Kyoto Protocol.* FCCC/KP/CMP/2005/8/Add.1.

Decision 3/CMP.1 (2005). *Modalities and Procedures for a Clean Development Mechanism as Defined in Article 12 of the Kyoto Protocol.* FCCC/KP/CMP/2005/8/Add.1.

Decision 5/CMP.1 (2005). *Modalities and Procedures for Afforestation and Reforestation Project Activities under the Clean Development Mechanism in the First Commitment Period of the Kyoto Protocol.* FCCC/KP/CMP/2005/8/Add.1.

Decision 12/CMP.1 (2005). *Guidance Relating to Registry Systems under Article 7, Paragraph 4, of the Kyoto Protocol.* FCCC/KP/CMP/2005/8/Add.2.

Decision 13/CMP.1 (2005). *Modalities for the Accounting of Assigned Amounts under Article 7, Paragraph 4, of the Kyoto Protocol.* FCCC/KP/CMP/2005/8/Add.2.

Decision 15/CMP.1 (2005). *Guidelines for the Preparation of the Information Required under Article 7 of the Kyoto Protocol.* FCCC/KP/CMP/2005/8/Add.2.

Decision 19/CMP.1 (2005). *Guidelines for National Systems under Article 5, Paragraph 1, of the Kyoto Protocol.* FCCC/KP/CMP/2005/8/Add.3.

Decision 20/CMP.1 (2005). *Good Practice Guidance and Adjustments under Article 5, Paragraph 2 of the Kyoto Protocol.* FCCC/KP/CMP/2005/8/Add.3.

Decision 22/CMP.1 (2005). *Guidelines for Review under Article 8 of the Kyoto Protocol.* FCCC/KP/CMP/2005/8/Add.3.

Decision 27/CMP.1 (2005). *Procedures and Mechanisms Relating to Compliance under the Kyoto Protocol.* FCCC/KP/CMP/2005/8/Add.3.

Decision 1/CMP.3 (2007). *Adaptation Fund.* FCCC/KP/CMP/2007/9/Add.1.

Decision 1/CMP.7 (2011). *Outcome of the Work of the Ad Hoc Working Group on Further Commitments for Annex I Parties under the Kyoto Protocol at Its Sixteenth Session.* FCCC/KP/CMP/2011/10/Add.1.

Decision 2/CMP.7 (2011). *Land Use, Land-Use Change and Forestry.* FCCC/KP/CMP/2011/10/Add.1.

Decision 8/CMP.7 (2011). *Further Guidance Relating to the Clean Development Mechanism.* FCCC/KP/CMP/2011/10/Add.2.

Decision 1/CMP.8 (2012). *Amendment to the Kyoto Protocol Pursuant to Its Article 3, Paragraph 9 (the Doha Amendment).* FCCC/KP/CMP/2012/13/Add.1.

Decision 1/CMP.9 (2013). *Report of the Adaptation Fund Board.* FCCC/KP/CMP/2013/9/Add.1.

Decision 3/CMP.9 (2013). *Guidance Relating to the Clean Development Mechanism.* FCCC/KP/CMP/2013/9/Add.1.

Decision 6/CMP.9 (2013). *Guidance for Reporting Information on Activities under Article 3, Paragraphs 3 and 4, of the Kyoto Protocol.* FCCC/KP/CMP/2013/9/Add.1.

Decision 8/CMP.9 (2013). *Compliance Committee.* FCCC/KP/CMP/2013/9/Add.1.

5. Reports of subordinate bodies to the FCCC and Kyoto Protocol

Ad Hoc Working Group on the Durban Platform for Enhanced Action (ADP) (2013). *Note by the Co-Chairs: Note on Progress.* ADP.2013.14.InformalNote.

Ad Hoc Working Group on the Durban Platform for Enhanced Action (ADP) (2014a). *Reflections on Progress Made at the Third Part of the Second Session of the Ad Hoc*

Working Group on the Durban Platform for Enhanced Action and on Its Work in 2014. ADP.2014.1.InformalNote.

Ad Hoc Working Group on the Durban Platform for Enhanced Action (ADP) (2014b). *Reflections on Progress Made at the Fourth Part of the Second Session of the Ad Hoc Working Group on the Durban Platform for Enhanced Action.* ADP.2014.3.InformalNote.

CDM Executive Board (2007). *Thirty-Fifth Meeting Report.* CDM-EB-35.

CDM Executive Board (2009). *Annual Report 2009.* FCCC/KP/CMP/2009/16.

CDM Executive Board (2011a). *Annual Report 2011.* FCCC/KP/CMP/2011/3 (Part I).

CDM Executive Board (2011b). *Benefits of the Clean Development Mechanism 2011.*

CDM Executive Board (2012a). *Annual Report 2012.* FCCC/KP/CMP/2012/3 (Part I).

CDM Executive Board (2012b). *Benefits of the Clean Development Mechanism 2012.*

CDM Executive Board (2013). *Annual Report 2013.* FCCC/KP/CMP/2013/5 (Part I).

Clean Development Mechanism (2012). *Tool for the Demonstration and Assessment of Additionality (v. 7.0).* CDM-EB70 Annex 8.

Clean Development Mechanism (2013a). *CDM Accreditation Standard (v. 5.1).* CDM-EB46-A02-STAN.

Clean Development Mechanism (2013b). *CDM Demonstration of Additionality, Development of Eligibility Criteria and Application of Multiple Methodologies for Programmes of Activities (v. 3.0).* CDM-EB65-A03-STAN.

Clean Development Mechanism (2013c). *CDM Project Cycle Procedure (v. 5.0).* CDM-EB65-A32-PROC.

Clean Development Mechanism (2013d). *CDM Project Standard (v. 5.0).* CDM-EB65-A05-STAN.

Clean Development Mechanism (2013e). *CDM Validation and Verification Standard (v. 5.0).* CDM-EB65-A04-STAN.

Clean Development Mechanism (2014). *Voluntary Tool for Describing Sustainable Development Co-Benefits of CDM Project Activities or Programmes of Activities (v. 1.1).* SD-TOOL01.

Compliance Committee (2008). *Annual Report.* FCCC/KP/CMP/2008/5.

Compliance Committee (2011). *Annual Report.* FCCC/KP/CMP/2011/5.

Compliance Committee (2012). *Annual Report.* FCCC/KP/CMP/2012/6.

Compliance Committee (2013). *Annual Report.* FCCC/KP/CMP/2013/3.

Compliance Committee Enforcement Branch (2011a). *Report on the Sixteenth Meeting.* CC/EB/16/2011/2.

Compliance Committee Enforcement Branch (2011b). *Report on the Seventeenth Meeting.* CC/EB/17/2011/2.

Compliance Committee Enforcement Branch (2012a). *Report on the Eighteenth Meeting.* CC/EB/18/2012/3.

Compliance Committee Enforcement Branch (2012b). *Report on the Nineteenth Meeting.* CC/EB/19/2012/2.

Compliance Committee Enforcement Branch (2012c). *Report on the Twentieth Meeting.* CC/EB/20/2012/2.

Compliance Committee Enforcement Branch (2012d). *Report on the Twenty-First Meeting.* CC/EB/21/2012/2.

Compliance Committee Enforcement Branch (2013a). *Report on the Twenty-Second Meeting.* CC/EB/22/2013/3.

Compliance Committee Enforcement Branch (2013b). *Report on the Twenty-Third Meeting.* CC/EB/23/2013/3.

Compliance Committee Facilitative Branch (2011). *Report on the Tenth Meeting*. CC/FB/10/2011/3.
Compliance Committee Facilitative Branch (2012a). *Report on the Eleventh Meeting*. CC/FB/11/2012/2.
Compliance Committee Facilitative Branch (2012b). *Report on the Twelfth Meeting*. CC/FB/12/2012/3.
Compliance Committee Facilitative Branch (2013a). *Report on the Thirteenth Meeting*. CC/FB/13/2013/2.
Compliance Committee Facilitative Branch (2013b). *Report on the Fourteenth Meeting*. CC/FB/14/2013/3.
Compliance Committee Plenary (2006). *Report on the First Meeting*. CC/1/2006/4.
Compliance Committee Plenary (2012). *Report on the Tenth Meeting*. CC/10/2012/2.
Compliance Committee Plenary (2013a). *Report on the Twelfth Meeting*. CC/12/2013/3.
Compliance Committee Plenary (2013b). *Report on the Thirteenth Meeting*. CC/13/2013/7.
Expert Review Team (2011). *Report of the In-Depth Review of the Fifth National Communication of Germany*. FCCC/IDR.5/DEU.
Expert Review Team (2012). *Report of the In-Depth Review of the Fifth National Communication of Australia*. FCCC/IDR.5/AUS.
Expert Review Team (2013). *Report of the Individual Review of the Annual Submission of Australia Submitted in 2012*. FCCC/ARR/2012/AUS.
FCCC Interim Secretariat (1994). *First Compilation and Synthesis of First National Communications from Annex I Parties*. A/AC.237/81.
FCCC Secretariat (1996). *Second Compilation and Synthesis of First National Communications from Annex I Parties: Executive Summary*. FCCC/CP/1996/12.
FCCC Secretariat (1998). *Second Compilation and Synthesis of Second National Communications from Annex I Parties*. FCCC/CP/1998/11.
FCCC Secretariat (2007). *Investment and Financial Flows to Address Climate Change*. Bonn: UNFCCC.
FCCC Secretariat (2008). *Investment and Financial Flows to Address Climate Change: An Update*. FCCC/TP/2008/7.
FCCC Secretariat (2011a). *Compilation of Economy-Wide Emission Reduction Targets to Be Implemented by Parties Included in Annex I to the Convention*. FCCC/SB/2011/INF.1/Rev.1.
FCCC Secretariat (2011b). *Compilation of Information on Nationally Appropriate Mitigation Actions to Be Implemented by Parties Not Included in Annex I to the Convention*. FCCC/AWGLCA/2011/INF.1.
FCCC Secretariat (2011c). *Procedural Requirements and the Scope and Content of Applicable Law for the Consideration of Appeals under Decision 27/CMP.1 and Other Relevant Decisions of the Conference of the Parties Serving as the Meeting of the Parties to the Kyoto Protocol, as Well as the Approach Taken by Other Relevant International Bodies Relating to Denial of Due Process*. FCCC/TP/2011/6.
FCCC Secretariat (2011d). *Submissions on Information from Developed Country Parties on the Resources Provided to Fulfil the Commitment Referred to in Decision 1/CP.16, Paragraph 95*. FCCC/CP/2011/INF.1.
FCCC Secretariat (2011e). *Withdrawal by Croatia of Its Appeal against a Final Decision of the Enforcement Branch of the Compliance Committee*. FCCC/KP/CMP/2011/2.
FCCC Secretariat (2013). *Quantified Economy-Wide Emission Reduction Targets by Developed Country Parties to the Convention: Assumptions, Conditions, Commonalities and Differences in Approaches and Comparison of the Level of Emission Reduction Efforts*. FCCC/TP/2013/7.

Green Climate Fund (2013). *Report of the Green Climate Fund to the Conference of the Parties*. FCCC/CP/2013/6.
Standing Committee on Finance (2013). *Report of the Standing Committee on Finance to the Conference of the Parties*. FCCC/CP/2013/8.
Subsidiary Body for Implementation (SBI) (1997). *First Compilation and Synthesis of Second National Communications from Annex I Parties*. FCCC/SBI/1997/19.
Subsidiary Body for Implementation (SBI) (1999). *First Compilation and Synthesis of Initial National Communications from non-Annex I Parties*. FCCC/SBI/1999/11.
Subsidiary Body for Implementation (SBI) (2000). *Second Compilation and Synthesis of Initial National Communications from non-Annex I Parties*. FCCC/SBI/2000/15.
Subsidiary Body for Implementation (SBI) (2001). *Third Compilation and Synthesis of Initial National Communications from non-Annex I Parties: Executive Summary*. FCCC/SBI/2001/14.
Subsidiary Body for Implementation (SBI) (2002). *Fourth Compilation and Synthesis of Initial National Communications from non-Annex I Parties: Executive Summary*. FCCC/SBI/2002/8.
Subsidiary Body for Implementation (SBI) (2003). *Compilation and Synthesis of Third National Communications from Annex I Parties: Executive Summary*. FCCC/SBI/2003/7.
Subsidiary Body for Implementation (SBI) (2003). *Fifth Compilation and Synthesis of Initial National Communications from non-Annex I Parties*. FCCC/SBI/2003/13.
Subsidiary Body for Implementation (SBI) (2005). *Sixth Compilation and Synthesis of Initial National Communications from Non-Annex I Parties; Addendum: Inventories of Anthropogenic Emissions by Sources and Removals by Sinks of Greenhouse Gases*. FCCC/SBI/2005/18/Add.2.
Subsidiary Body for Implementation (SBI) (2005). *Sixth Compilation and Synthesis of Initial National Communications from non-Annex I Parties: Executive Summary*. FCCC/SBI/2005/18.
Subsidiary Body for Implementation (SBI) (2007). *An Assessment of the Funding Necessary to Assist Developing Countries in Meeting Their Commitments Relating to the Global Environment Facility Replenishment Cycle*. FCCC/SBI/2007/21.
Subsidiary Body for Implementation (SBI) (2007). *Compilation and Synthesis of Fourth National Communications from Annex I Parties: Executive Summary*. FCCC/SBI/2007/INF.6.
Subsidiary Body for Implementation (SBI) (2009). *Synthesis of Experiences and Lessons Learned in the Use of Performance Indicators for Monitoring and Evaluating Capacity-Building at the National and Global Levels*. FCCC/SBI/2009/5.
Subsidiary Body for Implementation (SBI) (2011). *Compilation and Synthesis of Fifth National Communications from Annex I Parties: Executive Summary*. FCCC/SBI/2011/INF.1.
Subsidiary Body for Implementation (SBI) (2013). *Compilation of Information on Nationally Appropriate Mitigation Actions to be Implemented by Developing Country Parties*. FCCC/SBI/2013/INF.12/Rev.2.
Subsidiary Body for Scientific and Technical Advice (SBSTA) (2004). *Guidelines for the Preparation of National Communications by Parties Included in Annex I to the Convention, Part I: UNFCCC Reporting Guidelines on Annual Inventories (Following Incorporation of the Provisions of Decision 13/CP.9)*. FCCC/SBSTA/2004/8.
Subsidiary Body for Scientific and Technical Advice (SBSTA) (2006). *Updated UNFCCC Reporting Guidelines on Annual Inventories Following Incorporation of the Provisions of Decision 14/CP.11*. FCCC/SBSTA/2006/9.

Subsidiary Body for Scientific and Technical Advice (SBSTA) (2011). *Report on the Expert Meeting on Forest Reference Emission Levels and Forest Reference Levels for Implementation of REDD-Plus Activities*. FCCC/SBSTA/2011/INF.18.

Work Programme on Long-Term Finance (2013). *Report on the Outcomes of the Extended Work Programme on Long-Term Finance*. FCCC/CP/2013/7.

6. State government reports and documentation

Australian Government (2011). *Australia's Fast-Start Finance: Progress Report*. Canberra.

European Union (2009). *Decision No. 406/2009/EC of the European Parliament and of the Council of 23 April 2009 on the Effort of Member States to Reduce their Greenhouse Gas Emissions to Meet the Community's Greenhouse Gas Emission Reduction Commitments Up to 2020*.

Germany (Federal Ministry for the Environment, Nature Conservation, and Nuclear Safety) and United Kingdom (Department of Energy and Climate Change) (2013). *International NAMA Facility: General Information Document*.

United States (2014). *US Submission on Elements of the 2015 Agreement* (available from <http://unfccc.int/files/documentation/submissions_from_parties/adp/application/pdf/u.s._submission_on_elements_of_the_2105_agreement.pdf>).

United States, Department of State (2011). *Meeting the Fast Start Commitment: US Climate Finance in Fiscal Year 2011*.

Index

2°C limit 13, 87–90, 103, 107, 109–10, 120, 133–4, 168–9, 171, 173–4

accountable reporting rule 1–2, 6–7, 29, 62, 83–6, 99, 101, 119, 130, 135, 158–9, 163, 166–8, 172–3; and whether compliance system is necessary 83–4, 101–3, 167, 169–70, 172–3
adaptation 5–6, 35, 118–21, 124–5
Ad Hoc Working Group on the Durban Platform for Enhanced Action (ADP) 83, 100–1, 103, 111, 115; and compliance system in post-2020 agreement 103
Adaptation Fund 124–5
additionality *see* CDM
Alexeew, J. 151
Annex I/non-Annex I differentiation 12, 15, 33–5, 45–7, 49, 53, 88, 98, 101, 107–8, 115–17
Argentina, climate law in 114
Ascui, F. 162–3
assessment of state reports 30, 40–7; *see also* Expert Review Teams; greenhouse gas inventories; national communications
assigned amount 39, 73, 80, 92–3, 112–14
Australia: climate law in 5, 22–4, 111; fast-start finance 131–2; repeal of climate laws 23, 127; temperature extremes in 23; and transfer of finance 51–2
AWG-KP 99–101
AWG-LCA 100–1

Bali Action Plan 93, 96, 125, 154–5
biennial reports 43–7
Bodansky, D. 61
bottom-up versus top-down approaches to mitigation commitments 2, 45, 58, 89, 92–3, 96, 101–2, 126, 168–9, 171–3

Brunnée, J. 15, 85, 115
burden-sharing 13, 45, 90–1, 98–9, 103, 107, 110–11, 126, 168–9, 172
Bureau (of the Compliance Committee) 66, 72, 78

Canada: climate law in 21–2, 111; correspondence with Facilitative Branch 79; non-achievement of Kyoto Protocol target 78, 114, 170, 172; withdrawal from Kyoto Protocol 12, 79, 99
Cancun COP 91, 125, 127, 155–6
Carbon Market Watch 147, 150
China: emission growth 107; and international compliance-system design 83; mitigation action 3, 50
Clean Air Act (United States) 21
Clean Development Mechanism (CDM) 7, 36, 39, 92, 118, 135–52, 154, 173; and additionality 136–7, 139–43, 145–9, 173; compliance issues 145–52, 163–4, 167, 173; Designated National Authority 144, 150; Designated Operational Entity 137–40, 145–7; Executive Board 136–41, 143–4, 147–52, 163; HFC-23 projects 148; project crediting period 138, 146; and sustainable development 136–7, 140, 143–5, 149–52, 173
climate change law: anaemic condition of 1–2, 5, 7–10, 15–26, 173–4; defined 1, 5–6, 13–14, 18, 24; need for/value of 8, 27; normative basis *see* normative law; *see also* accountable reporting rule; general mitigation rule
climate-finance rule *see* financial-support rule
commitment period *see* first commitment period obligations; Kyoto Protocol, commitment periods; second commitment period obligations

Index

common but differentiated responsibilities, principle of 29
compliance, defined 25–6; *see also* compliance of states
Compliance Committee 12, 43, 63–4, 66–7, 74–6, 78, 83–4; discretion in decision-making 67, 73; evaluation of performance 74–85; member qualifications 67; plenary 71, 76; *see also* Bureau; Enforcement Branch; Facilitative Branch
compliance of states: with accountable reporting rule 7, 29, 47–57, 167, 169, 173; with financial-support rule 7, 129–34, 167, 169; with general mitigation rule 29, 107–14, 168–9, 171, 173–4; with specific mitigation rule 103–7, 170; with targets for first commitment period 112–15, 170; *see also* CDM; REDD
Copenhagen Accord 2, 91, 93, 123–5, 133
Copenhagen COP 155, 165
courts *see* litigation, climate-change
Couzens, E. 27, 113
customary international law 2, 15, 29, 86, 167–8

'dangerous' climate change, defined 13, 87, 168
deforestation 153–4
developed/developing country differentiation *see* Annex I/non-Annex I differentiation
Disch, D. 151
Doha Amendment (to Kyoto Protocol) 14, 99
domestic climate change law, inflated accounts of 5
Durban COP 91, 100, 127, 157

effort-sharing *see* burden-sharing
emission budget, global 6–7, 45, 99, 102–3, 108–10, 168, 172
emission gap 108, 110, 171–2
emission-reduction commitment types: absolute reduction 4–5, 96–8, 102; emission-intensity reduction 3–4, 96–8; reduction from business as usual 3–4, 96–8, 102
energy efficiency 3, 18, 20, 22, 26, 151
Enforcement Branch (of the Compliance Committee) 66, 68, 72–4, 80–2, 85, 93, 170; adjustment power 72–3, 80; *Bulgaria* case 77; case load 71, 77, 80, 82–3; discretion 73; facilitative role 81;

penalties/consequences 73–4, 93, 170; state representation at hearings 81
Environmental Investigation Agency 161–2
equitable burden-sharing *see* burden-sharing
EU ETS 138, 140
European Union: climate law in 5, 18–19; and international compliance-system design 83
Expert Review Teams 40–3, 51–2, 54–5, 63–5, 69, 71, 73, 79–80, 92, 130, 169–70; constitution 40; facilitative nature 64, 74–7, 80–3; limited review powers 55–7, 74; and mandatory language in guidelines 64–6, 76–7; tendency to withhold questions of implementation from Compliance Committee 75–7; whether truly independent 52, 75–6; *see also* assessment of state reports

Facilitative Branch (of the Compliance Committee) 12, 66–72, 76–7, 170; and Canada 78–9; early-warning function 68, 71, 77–9; and Italy 78; measures/consequences 68–9; performance issues 70–1, 76–80, 83; and South Africa's referral 78
fast-start finance 125–6, 128, 131–3
Faure, M.G. 61
FCCC: 1990–2000 period obligations 12, 88–9; 2001–2007 period obligations 12, 92; 2013–2020 period obligations 3, 12, 15, 43–7, 62, 91, 93–9, 115, 158, 169, 171; emission-reduction obligations 86–91; lack of compliance system 90, 96, 99
Fifth Assessment Report of the IPCC *see* IPCC Fifth Assessment Report
finance, transfer of 6–7, 30, 33–4, 36, 44–5, 48, 51–2, 88–90, 115, 118–34
financial-support rule 6–7, 118–21, 134–5, 167, 169; *see also* general mitigation rule
first commitment period obligations 91–3, 112–15, 170
flexibility mechanisms 35–6, 92, 136; and participation eligibility 72–3, 81–2
fossil fuels, legal and economic support for 16–18, 26
France, climate law in 20
Fransen, T. 52, 55

general mitigation rule 2, 6–7, 13, 86–92, 96, 98–9, 101, 103, 115, 118–19,

129, 135, 159, 166, 168, 171, 173–4; communal and individualistic aspects 90–1, 103, 107–8, 111, 166, 168; *see also* compliance of states
Germany, fast-start finance 132
good-faith obligation 90–1, 99, 107–8, 111, 168
global emission budget *see* emission budget, global
Global Environment Facility 34, 123, 128–30
GLOBE climate law survey 18
Green Climate Fund 123, 128–9
greenhouse gas inventories: adjustment (by Enforcement Branch) 72–3, 80; content 30, 37–9; expert review 41–3; mandatory versus discretionary elements in guidelines 38, 41–2, 64–6; state compliance with reporting requirements 53–7; submission frequency 30, 37, 53–4; uncertainty in 55–61
greenhouse gas emissions: allowance trading 18, 20, 35–6, 93; growth of 109; from land use and forestry (or LULUCF) 39, 53, 57, 72, 103, 105, 112–14, 153, 160; as pollution 15–16, 22; pricing 16–19, 22–3, 127

High-Level Advisory Group on Climate Change Financing 127
Hilson, C. 21

India, climate law in 114
International Assessment and Review (IAR) 12, 44–5, 47, 61–2, 83, 99, 167, 172
International Consultation and Analysis (ICA) 12, 44, 46–7, 51, 61–2, 83, 99, 167, 172
International Energy Agency (IEA) 126
International Law Association 2
IPCC Fifth Assessment Report 1, 8–10, 109, 147
IPCC RCP2.6 scenario 109–10

Japan: climate law in 20–1; fast-start finance 132; non-participation in Kyoto Protocol's second commitment period 12, 21, 99
Joint Implementation 92

Kenya, and REDD 162
Keystone XL Pipeline 16
Kidd, M. 27, 113
Kulovesi, K. 131, 133–4

Kyoto Protocol: being only a minor advance on FCCC 2, 92, 115, 168; commitment periods 12, 14, 39, 74, 91–3, 99–100; compliance system *see* Compliance Committee; disillusionment with 14–15; quantified emission limitation and reduction commitment 36, 72, 92 99; *see also* assigned amount

Laos, illegal logging 161–2
Least Developed Countries Fund 124
Lefeber, R. 70, 78
Lefevere, J. 61
Lin, J. 163–4
litigation, climate-change: political nature of 24; as source of law 24–5
long-term finance 125–8, 130–1, 133–4
Lund, E. 140

managerialism 84
Massachusetts v. EPA 25
Mehling, M. 61–2
mitigation *see* emission-reduction commitment types; financial-support rule; general mitigation rule; specific mitigation rule
Mundaca, L. 20
Nakhooda, S. 132–3

NAMA *see* nationally appropriate mitigation action
national communications: content 30–7; expert review 40–3; mandatory versus discretionary elements in guidelines 31–3, 64–6; state compliance with reporting requirements 48–52; submission frequency 28–9, 35, 43–4, 48–9
national inventory report (NIR) 38–9
nationally appropriate mitigation action (NAMA) 3, 93, 96–8, 106, 125, 129
national registry 36, 39, 72, 80
National Research Council, United States 59
national system 12, 35, 43, 65, 68, 72, 80
Neeff, T. 162–3
Nepal, and REDD 162
New Zealand: climate law in 20; non-participation in Kyoto Protocol's second commitment period 12, 99; and self-determined mitigation targets 102
normative law: generally 10–11; in climate change law 13–25, 85, 89–90, 120, 135, 166, 168
Norway, fast-start finance 132
nuclear energy 20

Oberthür, S. 47
OCO-2 satellite 59
Olbrisch, S. 130–1
'optional' mitigation mechanisms *see* CDM; REDD

Paris agreement *see* post-2020 agreement
per-capita emissions 115–17
pledge period *see* FCCC 2013–2020 period obligations
population growth 116–17
post-2020 agreement 2, 6, 62, 83, 99–103, 133–4, 173; and nationally determined contributions 101
'potential problem' *see* questions of implementation
prevention rule *see* general mitigation rule

questions of implementation 43, 63–7, 74–7; and the Enforcement Branch 68, 72–3; and the Facilitative Branch 68–9, 76–80

Rajamani, L. 47, 62, 145
Reducing Emissions from Deforestation and forest Degradation (REDD) 7, 135, 152–64, 173; compliance issues 158, 160–4, 167, 173; finance for 157–9, 161–3, 173; safeguards 156
Reid, C.T. 19
renewable energy 18–22, 26, 114, 131
reporting, by states *see* greenhouse gas inventories; national communications
reporting rule *see* accountable reporting rule
review of state reports *see* assessment of state reports; Expert Review Teams; greenhouse gas inventories; national communications
Richter, J.L. 20
Romijn, E. 160–1
Roster of Experts 46, 81, 158
Russia: climate law in 111; deforestation 154; non-participation in Kyoto Protocol's second commitment period 12, 99

Schneider, L. 147, 150–1
second commitment period obligations 92, 95, 99–100
Secretariat, FCCC 28, 40, 47–51, 53, 90, 103–5, 110–11, 126
South Africa: climate law in 27, 113; and international compliance-system design 102; referral of question of implementation 78; and top-down mitigation targets 102
Special Climate Change Fund 124
specific mitigation rule 14, 86–92, 96, 99, 115, 170
state compliance *see* compliance of states
Streck, C. 163–4
Subsidiary Body for Scientific and Technological Advice (SBSTA) 93–5, 122, 155
Subsidiary Body on Implementation (SBI) 41, 45–6, 90, 96, 103–7
sustainable development *see* CDM
Switzerland: climate law in 20; and international compliance-system design 102

team of technical experts (TTE) 46
technology transfer 30, 33, 36, 69, 96, 121, 123–4
Toope, S.J. 15, 85
top-down versus bottom-up *see* bottom-up versus top-down approaches to mitigation commitments *or* top-down verification of emissions
top-down verification of emissions 57–61
transparency rule *see* accountable reporting rule

Ulfstein, G. 55
United Kingdom: climate law in 5, 19; fast-start finance 132
United Nations Framework Convention on Climate Change *see* FCCC
United States: climate law in 16, 21–2, 111–13; fast-start finance 131–2; and international compliance-system design 83–4, 103, 172; and long-term finance 133–4; non-participation in Kyoto Protocol 12; pledge for the 2012–2013 period 96; and self-determined mitigation targets 101–2

Vietnam, illegal logging 161–2
Vihma, A. 47

Werksman, J. 55
Wiersema, A. 162
Work Programme on Long-Term Finance 133–4

Yamineva, Y. 131, 133–4

eBooks
from Taylor & Francis
Helping you to choose the right eBooks for your Library

Add to your library's digital collection today with Taylor & Francis eBooks. We have over 50,000 eBooks in the Humanities, Social Sciences, Behavioural Sciences, Built Environment and Law, from leading imprints, including Routledge, Focal Press and Psychology Press.

Choose from a range of subject packages or create your own!

Benefits for you
- Free MARC records
- COUNTER-compliant usage statistics
- Flexible purchase and pricing options
- 70% approx of our eBooks are now DRM-free.

Benefits for your user
- Off-site, anytime access via Athens or referring URL
- Print or copy pages or chapters
- Full content search
- Bookmark, highlight and annotate text
- Access to thousands of pages of quality research at the click of a button.

 **Free Trials Available**

We offer free trials to qualifying academic, corporate and government customers.

eCollections
Choose from 20 different subject eCollections, including:
- Asian Studies
- Economics
- Health Studies
- Law
- Middle East Studies

eFocus
We have 16 cutting-edge interdisciplinary collections, including:
- Development Studies
- The Environment
- Islam
- Korea
- Urban Studies

For more information, pricing enquiries or to order a free trial, please contact your local sales team:

UK/Rest of World: **online.sales@tandf.co.uk**
USA/Canada/Latin America: **e-reference@taylorandfrancis.com**
East/Southeast Asia: **martin.jack@tandf.com.sg**
India: **journalsales@tandfindia.com**

www.tandfebooks.com